Applications of Data Science to Molecules and Materials 2021

Topical Conference at the 2021 AIChE Annual Meeting

Boston, Massachusetts, USA and Online
7 - 11 November and 15 - 19 November 2021

ISBN: 978-1-7138-5689-4

Printed from e-media with permission by:

Curran Associates, Inc.
57 Morehouse Lane
Red Hook, NY 12571

Some format issues inherent in the e-media version may also appear in this print version.

Printed with permission by Curran Associates, Inc. (2022)

For permission requests, please contact AIChE
at the address below.

AIChE
120 Wall Street, FL 23
New York, NY 10005-4020

Phone: (800) 242-4363
Fax: (203) 775-5177

www.aiche.org

Additional copies of this publication are available from:

Curran Associates, Inc.
57 Morehouse Lane
Red Hook, NY 12571 USA
Phone: 845-758-0400
Fax: 845-758-2633
Email: curran@proceedings.com
Web: www.proceedings.com

TABLE OF CONTENTS

624936 Understanding Barrierless Adsorption in Electro-Catalytic Systems Using Accelerated Transition Path Sampling

Abstract:

The adsorption and desorption processes that often initiate or terminate catalytic reaction sequences are challenging to account for in kinetic models, since they are often barrierless processes where conventional transition-state theory (TST) is not applicable. In particular, the adsorption or desorption of carbon monoxide (CO) on metal surfaces is a very important process as it is a rate limiting step for many catalytic reactions. The free energy profiles of these processes are even more difficult to calculate when there is solvent present, because the potential energy surface (PES) will be rugged and fluctuating. One approach to studying the kinetics and thermodynamics of a process such as this is to employ a rare event sampling method such as transition path sampling (TPS) which requires molecular dynamics simulations to generate dynamical trajectories of all the atoms in the system. However, molecular dynamics using density functional theory (DFT) is highly computationally expensive and not practical. Data-driven machine learning force fields exploit machine learning methods to quickly predict forces and energies in atomistic simulations. We use classical force-fields as well as machine-learned potentials to carry out TPS to calculate free energy profiles of adsorption/desorption of CO on Pt(111). We use the Atomistic Machine-learning Package (*Amp*) to train a high dimensional neural network potential on the reference data calculated using DFT GGA (BEEF-vdW) with finite-difference mode. First, we perform TPS calculations for desorption of CO at solid-liquid interface to identify all of the relevant reaction coordinates. Then we use meta-dynamics simulations to calculate free-energy profile across multivariable as reaction coordinates and compare results with umbrella sampling calculations done along one reaction coordinate. We also compare the CO desorption at solid-liquid interface with solid-gas interface to quantify the effect of solvents on the dynamics of the desorption process. This approach can be easily generalized to other machine-learning potentials and opens up a path to study kinetics of systems with non-smooth PES.

627253 Methods and Applications of *Ex-Machina* Molecular Dynamics to Catalytic Processes

Abstract:

We pursue the paradigm of "ex-machina" computations where data-driven approximations are automatically developed using machine learning algorithms and enable access to dynamics of complex chemical systems. Non-parametric regression methods allow for learning of potential energy surfaces from expensive quantum calculations. To accelerate molecular dynamics calculations, we developed the Neural equivariant interatomic potential model (NequIP) based on tensor-valued symmetry-preserving layer architectures and used them to achieve state-of-the-art accuracy and training efficiency for simulating dynamics of molecules, water and heterogeneous catalysts. In order to enable autonomous selection of the training set for reactive systems, we developed the FLARE adaptive closed-loop algorithm that constructs accurate and uncertainty-aware Bayesian force fields on-the-fly from a molecular dynamics simulation, using Gaussian process regression. We demonstrate the performance of ML-accelerated MD simulations by studying bimetallic catalyst surface restructuring and HD exchange reactions. Finally, we develop dimensionality reduction techniques in order to automatically identify the reaction coordinates from dynamics simulations, that can be used to enhance sampling of rare transitions and to estimate reaction rates.

623636 Statistical Models for Predicting Oil Composition from Hydrothermal Liquefaction of Biomass

Abstract:

Hydrothermal liquefaction (HTL) converts wet biomass to energy-rich bio-oil to serve as a renewable substitute to crude oil derived from fossil fuels. HTL uses hot pressurized water to break down biomass and undergo a series of reactions (hydrolysis, de-carboxylation, re-combination) to form a large number of small organic molecules. The properties of crude oil are closely related to the chemical composition of these small organic molecules. Hence, there is an increased interest within the waste valorization community to analyze the composition of the oil and further predict the oil composition at any given process condition (temperature, time) for any feedstock. There is also an opportunity to further optimize the process - to increase the oil yield obtained using this knowledge.

Biomass is a diverse feedstock composed of several types of polysaccharides, proteins, lignin, and lipids. Moreover, these components also interact with each other to alter oil compositions. Due to this, developing compositional models that are generic to a wide variety of feedstocks has proven challenging. Also, since HTL is a high temperature – high-pressure process, performing high throughput experimentation is difficult. Hence, we utilize data mined from HTL literature and statistical methods (multivariate linear regression, regression tree, random forest) to predict oil compositions.

The composition of HTL oil is mostly characterized using GC/MS. This information is predominantly classified into product classes (esters, long-chain fatty acids, hydrocarbons) and given in bar charts in literature. We developed an in-house ChartReader model that automatically parses data from bar charts in literature. Using the ChartReader model and human inspection, a data set with 682 data points is prepared. Each of these data points contains oil composition (in functional classes), feedstock composition (biochemical and elemental), and process variable (temperature, time). The products formed in the oil phase are lumped into 9 product classes: esters, oxygenated single ring aromatics, furans, long chain fatty acids, long chain alcohols, aldehydes, and ketones, N-containing compounds, aliphatics, and polycyclic aromatics. Note that the sum of all product class percentages is 100%.

For modeling the oil composition, the biochemical/ elemental compositions (and all two-way interactions involved), temperature, and time are considered the full set of predictors. In cases where the biochemical or elemental composition was missing, the EM algorithm was used to impute this information from the other. The response (oil composition) was represented by percentages for the 9 classes. Using MANOVA in the context of multivariate multiple linear regression (MLR), important predictors were identified. The same predictors were also used to fit regression tree (RT) random forest (RF) models. The MLR, RT, and RF models were subsequently compared using various measures of prediction error.

Our initial analysis indicates, MLR model predicting all product classes with $R^2 > 0.8$ (sum of squared residual between experimental and predicted values). The mean absolute error and RMSE of all product classes are 11.6 and 15.9, respectively. All binary interactional effects except for protein-lipid interactions are significant in influencing oil composition. The MAE and RMSE values follow the following trend: MLR> RT> RF. Hence, RF predicts best among the models tested.

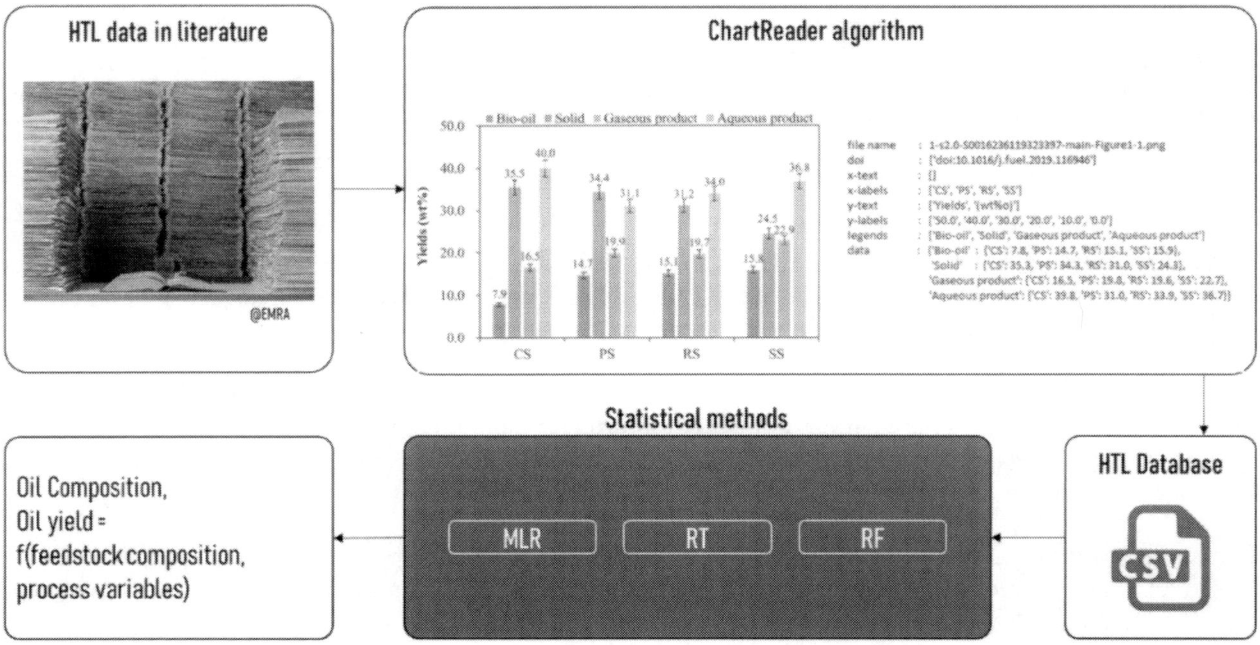

Fig: Overview of data mining approach adopted for developing predictive models for oil composition from hydrothermal liquefaction of biomass (MLR: multivariate linear regression, RT: regression tree, RF: random forest)

627632 A Big Data Approach to Predict the Kinetics of Organic Chemistry Reactions

Abstract:

Computing reaction rate constants ab initio quickly becomes unfeasible as the system size grows. In this talk we will discuss our ongoing work which aims to explore and employ machine learning algorithms to predict reactivity in the ground state. Our current efforts focus on creating a large reaction rate constant dataset for organic chemistry reactions to train supervised machine learning algorithms to predict quantum rate constants. The dataset is generated by extending our previous model dataset of over 1.5 million examples [1] to a set of known classes of chemical reactions such as e.g. nuclear substitution. Furthermore, dataset examples now include reactant partition functions computed using the rigid body and harmonic oscillator approximation at the DFT level of theory. With this dataset we are currently training ML algorithms to predict quantum reaction rate constants at various temperatures. In the future, we plan to extend this work further to include features to describe reactivity on catalytic surfaces.

[1] E. Komp and S. Valleau, "Machine Learning Quantum Reaction Rate Constants," *J. Phys. Chem. A*, 124:8607-8613, 2020.

628222 Uncertainty Quantification of Catalyst Structure Effects on Kinetics

Abstract:

Heterogeneous catalytic reactions are commonly structure sensitive and dependent on the size and shape of catalyst particles. However, the catalyst's structure effects on reactivity are often neglected in state-of-art microkinetic modeling by assuming the reaction happens only on a single crystallographic plane [1]. This material gap in identifying and modeling the real active site intrinsically hinders the understanding of complex reaction mechanisms at the atomic level. Published experimental data have long served as a valuable resource for bridging this gap. However, the reported literature data often exhibit significant differences for even identical systems. This fact complicates meaningful comparisons across sources and hinders the utilization of prior knowledge. To alleviate this problem, it is important to quantify the uncertainties caused by catalyst structure effects.

Complete methane oxidation on noble metals is chosen as a case study due to many available published literature data. The structure sensitivity of this reaction has been debated experimentally for decades [2] due to the difficulty in identifying catalyst surface structures in situ. Structure-dependent kinetic modeling provides the possibility to explore it theoretically. Recently published computational work either chooses several common facets [3] or develops a structure-dependent model for archetypical simple reactions, such as CO oxidation [4]. There is a need to create suitable microkinetic models for describing structure sensitivity and explore structure effects.

In this work, we develop a novel methodology to build a structure-descriptor-based kinetic model for complete methane oxidation based on first-principles calculations. Structure-reactivity scaling relationships, reminiscent of the generalized coordination number (GCN) [5], are developed using machine learning techniques. By incorporating these correlations into kinetic models leveraging our in-house developed software, we predict the experimental observables, such as turnover frequencies (TOF) and apparent activation energies, at a dramatically reduced computational cost compared to first-principles calculations. Additionally, uncertainty quantification is applied for exploring the effects of errors in structure-reactivity scaling relations on variable catalyst facets. This methodology enables the rapid prediction of kinetics and quantifies the uncertainties due to the catalyst structure.

References

1. Cheula, R., Soon, A., and Maestri, M. Catal. Sci. Technol. 8, 3493 (2018).

2. Beck, Irene E. et al. J. Catal. 268, 60 (2009).

3. Wang, Y. et al. J. Phys. Chem. C 124, 2501 (2020).

4. Jørgensen, M. and Grönbeck, H. ACS Catal. 7, 5054 (2017).

5. Calle-Vallejo, F., Sautet, P. and Loffreda, D., Angew. Chemie - Int. Ed. 53, 8316 (2014).

631720 Open Catalyst Project: Advancements and Challenges in Building a Flexible Machine Learning Potential

Abstract:

Machine learning potentials have shown considerable success in accelerating quantum mechanical calculations for computational catalysis applications. These applications are often focused on a particular chemistry or narrow combination of elements, limiting their ability to screen new catalysts. The Open Catalyst Project aims to develop new ML methods and models to accelerate the catalyst simulation process for renewable energy technologies and improve our ability to predict properties across catalyst composition. The initial release of the Open Catalyst 2020 (OC20) dataset presented the largest open dataset of thermochemical intermediates across multi-metallic surfaces, spanning 55 unique elements and 80 reaction intermediates. Since then, the development of state-of-the-art models and methods has been well underway by ourselves, collaborators, and members of both the catalysis and ML communities. I will introduce the Open Catalyst Project, the associated OC20 dataset, and recent updates since our initial release. Specifically I will discuss some of the progress in building state-of-the-art models and common challenges faced along the way.

626192 Machine Learning Accelerated Scale-up for Microporous Materials

Abstract:

Microporous materials such as zeolites and MOFs play a crucial role in producing energy and energy products at scale. Traditional approaches for materials development and scale-up are time consuming and involve a lot of trial and error. Two key questions need to be addressed including (1) what the critical variables are that impact the synthesis (2) and how the material properties can be controlled and optimized. This presentation provides an overview of machine learning approaches to build quantitative synthesis-property relationships (QSPR). These methods provide a highly efficient path to optimize synthesis parameters towards target(s) such as purity, crystal size and surface area, and enable us to enable us to significantly speed up our materials workflow.

Our workflow combines design of experiments, machine learning and deep learning, and high-throughput experimentation (HTE). In order to build QSPR, we featurized the characterization data using machine learning and deep learning approaches. For example, we quantified crystal purity using peak deconvolution of powder XRD pattern. We used a deep learning model to calculate crystal size and aspect ratio from scanning electron microscopy (SEM). We also performed functional principal component analysis to ensure the surface area calculated from the linear region of Brunauer-Emmett-Teller (BET) adsorption curve selected using Rouquerol rule explains a substantial fraction of variance. Since the synthesis parameter space for microporous materials is large and complex, we combined Bayesian Optimization and HTE to further accelerate the workflow. After optimization, we used feed-forward neural network models to summarize QSPR for extended investigation at different scales.

We validated the accelerated workflow with a known zeolite. Without referring to historical data, we used the workflow to systematically probe a large and complex synthesis parameter space and obtain small pure crystals of the material. The new workflow demonstrated a significant reduction in the number of experiments needed to meet the same goals as past experiments.

628298 Physics-Informed Neural Networks for Kinetic Parameter Estimation and Uncertainty Quantification

Abstract:

Top-down kinetics has mainly relied on the laborious acquisition and assessment of steady-state kinetic data obtained through standard design-of-experiments. In this work, we demonstrate that kinetic parameters can be retrieved from a reduced number of transient experiments given prior knowledge of the underlying mechanism and its associated kinetic model. Furthermore, we outline the structural limitation for information retrieval from surface experiments latent dynamics based on an algebraic indeterminacy criterion governed by stoichiometry. We utilize the mean-field approximation as a general surrogate for chemical kinetics, and artificial neural networks (ANNs) as basis functions for the solution of time-dependent ordinary differential equations (ODEs) by adapting the physics-informed artificial neural network (PINN) approach [1,2] to chemical systems' dynamics. Based on the PINN structure, we formulate the inverse chemical kinetics problem as physics-regularized regression problems (or data-regularized solution of ODEs) [3], and we extend the analysis to maximum-likelihood estimators (MLEs). MLEs provide the statistical structure for the estimation of confidence intervals for regressed model parameters through Hessian-based uncertainty quantification. This neural network-assisted framework for the analysis of chemical kinetics ODEs can be employed in the estimation of kinetic parameters and preliminary reaction mechanism discrimination based on transient data.

[1] Lagaris, I. E., Likas, A., & Fotiadis, D. I. (1997). Artificial Neural Networks for Solving Ordinary and Partial Differential Equations. IEEE Transactions on Neural Networks, 9(5), 987–1000.

[2] Raissi, M., Perdikaris, P., & Karniadakis, G. E. (2019). Physics-informed neural networks: A deep learning framework for solving forward and inverse problems involving nonlinear partial differential equations. Journal of Computational Physics, 378, 686–707.

[3] Gusmão, G. S., Retnanto, A. P., da Cunha, S. C. & Medford, A. J. (2020). Kinetics-Informed Neural Networks. ArXiv ID: 2011.14473

625853 Deep Learning Quantum Reaction Rate Constants

Abstract:

In this talk we will discuss our recent effort [1] to overcome the cost of ab initio reaction rate constant calculations by using supervised machine learning. A deep neural network, DNN, was trained on our in-house generated ~1.5 million example dataset of quantum reaction rate constants. Rate constants multiplied by reactant partition functions were computed for single, double, symmetric and asymmetric one-dimensional potentials over a broad range of reactant masses and temperatures. The network was trained for hundreds of epochs and able to predict the logarithm of the rate product with a relative mean absolute error of 1.1%. Systems beyond the test set were also studied, these included the diffusion of hydrogen on Ni(100), the Menshutkin reaction of pyridine with CH_3Br in the gas phase, the reaction of formalcyanohydrin with HS^- in water and the F + HCl reaction. The DNN predictions were in good agreement with the exact rates.

While this was encouraging, when exploring the test set predictions, some had errors as high as 33.5%. It was thus clear that anticipating individual prediction errors was necessary to inform design choices. We will discuss our recent effort to estimate that error using modified generative adversarial networks (GANs)[2-3].

[1] E. Komp and S. Valleau, "Machine Learning Quantum Reaction Rate Constants," *J. Phys. Chem. A*, 124:8607-8613, 2020.

[2] I. Goodfellow *et al.*, "Generative adversarial networks," *arXiv*:1406.2661, 2014.

[3] M. Lee and J. Seok, "Estimation with Uncertainty via Conditional Generative Adversarial Networks," *arXiv:*2007.00334 2020.

639149 Building Blocks for Autonomous Computing Materials: Dimers, Trimers and Tetramers

Abstract:

Autonomous computing materials for data storage and computing offer an opportunity for next generation of computing devices. Patchy nanoparticle networks, for example, have been suggested as potential candidates for emulating neuronal networks and performing brain-like computing. Here, we use molecular dynamics (MD) simulations to show that stable dimers, trimers and tetramers can be built from citrate capped gold nanoparticles (cit-AuNPs) linked by poly(allyamine hydrochloride) (PAH) chains. We use different lengths of PAHs to build patchy nanoparticle networks that can emulate a complex neuronal network linked by axons of varying lengths. We find that the tetramer structure can accommodate up to 11 different states when the AuNP pairs are connected by either of two polymer linkers, PAH200 and PAH300. We find that the heavy AuNPs are contributing to the assembly's structure stability. To further illustrate the stability, the AuNP-AuNP distances in dimer, trimer and tetramer structures are reduced by steering the cit-AuNPs closer to each other. At different distances, these steered structures are all locally stable in a 10 ns MD simulation time scale because of their connection to the AuNPs. We also find that the global potential energy minimum is at short AuNP-AuNP distances where AuNPs collapse, because the $-NH_3^+$ and $-COO^-$ attraction reduces the potential energy. The stability and application of these fundamental structures remain to be further improved through the use of alternative polymer linkers and nanoparticles.

622639 A Novel Machine Learning Model for the Accurate Design of Highly Selective Zeolitic-Imidazolate Frameworks

Abstract:

Zeolitic-imidazolate frameworks (ZIFs) have the potential to make highly selective membranes, due to their functionalization capabilities that can take place on the molecular level and affect considerably their microscopic properties. However, their application in gas separations is limited due to the lack of knowledge on the modification-properties correlation.

Our approach is based on the premise that the aperture connecting the cages of ZIFs controls the diffusion rate of guest species. In our recent works, we showed that the aperture's size and flexibility can change, by replacing the organic linker, the metal center or the functional group of the framework. This in turn provides control over the diffusion-based selectivity of a given pair.[1,2,3] This knowledge shapes a chemical basis which we used in a novel machine learning (ML) tool towards developing a complete modification/selectivity correlation for ZIFs, as shown schematically in Figure 1. We built numerous new ZIFs of finely discretized aperture sizes, by replacing various units, and simulated the diffusion of guest molecules of varying in size (He, H_2 up to n-butane) in these structures, based on force fields developed based on density functional theory (DFT) calculations. The simulations account for the flexibility of the structure, to assess the framework's response as a function of the gas penetrant's size. Then we trained a supervised ML model on descriptors with physical (size and weight of the linkers; charge and length of the functional group) and chemical significance (ionic radius of the metal center and information on the metal-nitrogen bond), as well as information on the response of the framework upon a given gas penetrant's presence.

Overall, our ML tool enables the prediction of the diffusivity of any gas on any ZIF structure and of the diffusivity of any gas in it without the need for molecular level simulations. The only information needed is basic chemical information from DFT calculations and the size of the gas penetrant. This approach can further promote the fast design of process tailored new ZIFs, by providing as input information only the desired diffusivity/selectivity of a given mixture.

1. Krokidas, P. *et al.* On the Efficient Separation of Gas Mixtures with the Mixed-Linker Zeolitic-Imidazolate Framework-7-8. *ACS Appl. Mater. Interfaces* **10**, 39631–39644 (2018).

2. Krokidas, P., Moncho, S., Brothers, E. N., Castier, M. & Economou, I. G. Tailoring the gas separation efficiency of metal organic framework ZIF-8 through metal substitution: A computational study. *Phys. Chem. Chem. Phys.* **20**, 4879–4892 (2018).

3. Krokidas, P., Moncho, S., Brothers, E. N. & Economou, I. G. Defining New Limits in Gas Separations Using Modified ZIF Systems. *ACS Appl. Mater. Interfaces* **12**, 20536–20547 (2020).

Figure 1. Outline of our ML model for the fast design of top-performing ZIFs with desired performance.

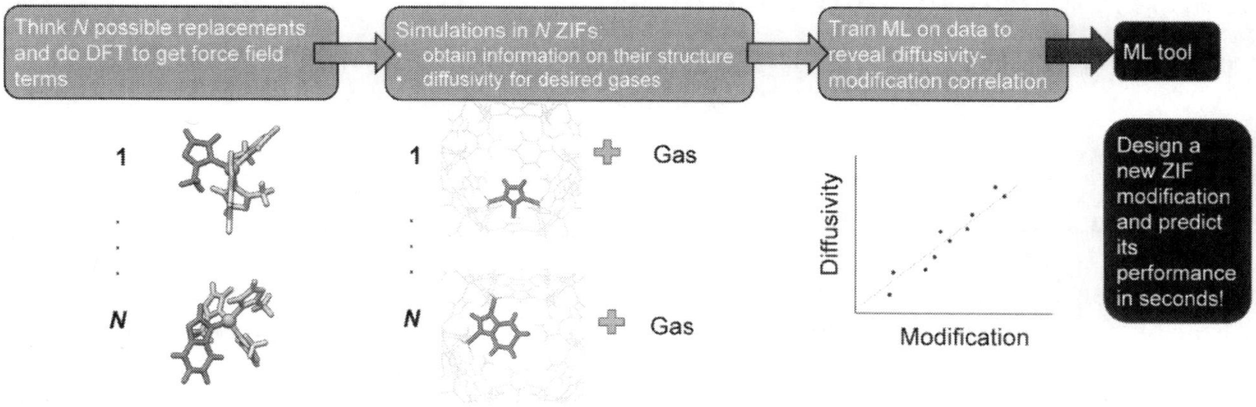
Think N possible replacements and do DFT to get force field terms
Simulations in N ZIFs:
• obtain information on their structure
• diffusivity for desired gases
Train ML on data to reveal diffusivity-modification correlation
ML tool
1
N
Gas
Diffusivity
Modification
Design a new ZIF modification and predict its performance in seconds!

624376 Designing and Synthesizing Novel Dye Molecules Using Generative Modeling and Data-Driven Synthesis Planning

Abstract:

Development of new products often relies on the discovery of novel molecules. While traditional molecular design involves using expert chemical knowledge to propose, synthesize, and test new molecules, this process can be cost and time expensive, limiting the number of molecules that can be reasonably tested. Recently, we have developed a generative modeling framework that is capable of producing novel molecules that are optimized with respect to multiple objectives or constraints. In this work, we combined this method with our previously developed, open-source synthesis-planning tool, ASKCOS, to generate novel dye molecules and their synthetic routes. We chose to focus on dye design because dyes have rich chemical diversity and can be used for a range of important applications included use in LED displays, functional coatings, or even biological labels for fundamental biological studies. We focused on the problem of designing a set of novel dyes that have an absorbance maximum at a range of target wavelengths. First, the generative modeling framework was used to generate a large number of dye molecules, with focus on producing dyes that absorbed at higher wavelengths (for which there are fewer examples in the literature). Then, top candidates for each target wavelength were screened for their synthetic accessibility (based on a synthetic accessibility score) and their novelty (based on comparisons with general and dye-specific datasets). ASKCOS was used to design synthetic routes for all generated molecules that were 1) predicted to absorb light at or near a target wavelength, 2) found to be synthetically accessible, and 3) found to be novel. Finally, a high throughput synthesis and testing platform was used to synthesize and verify that each dye molecule absorbed maximally near its target wavelength. This work illustrates how generative models and synthesis planning tools can be used together to discover new molecules and represents an important step toward achieving closed loop molecular discovery.

622110 Predicting Critical Micelle Concentrations for Surfactants Using Graph Convolutional Neural Networks

Abstract:

Surfactants are amphiphilic molecules that are widely used in consumer products [1], industrial processes [2], and biological research [3,4]. A critical property of a surfactant is the critical micelle concentration (CMC), at which surfactant molecules undergo cooperative self-assembly in solution [5]. The formation of micelles in surfactant solution can bring significant changes in properties such as surface tension, electrical conductivity, light scattering, and reaction mechanism [5,6]. Consequently, predicting CMCs is important for surfactant selection and design in applications such as drug delivery [7] and detergents [8].

Traditionally, CMCs are determined experimentally through techniques such as tensiometry, but this method is laborious and expensive [9]. As an alternative to experiments, computational methods such as molecular dynamics (MD) simulations [10, 11, 12] and quantitative structure-property relationships (QSPR) that are based on molecular descriptors [13, 14, 15] have been used to predict CMCs. These approaches have been shown to predict CMCs with relatively high accuracy, but they have several limitations. For instance, MD simulations usually require high computational cost and assumptions regarding the number of surfactants within a micelle [10, 11, 12], whereas QSPR models are often applicable to only a single class of surfactant and may need density functional theory (DFT) calculations to obtain quantum-chemical descriptors [16].

Recent advances in machine learning methods for molecular property prediction can help overcome some of these obstacles, specifically graph convolutional neural networks (GCNs) [17], since chemicals can be intuitively represented as molecular graphs that are natural inputs for GCNs. A GCN architecture can handle input molecules with different sizes without the need for artificial data manipulations and approximates the physical interactions between atoms through graph convolutions that aggregate features of adjacent atoms [17].

In this work, we present a simple GCN that can predict CMCs directly from the molecular structure of a surfactant monomer and test the ability of the model to generalize to a dataset containing nonionic, anionic, cationic, and zwitterionic surfactants [5]. We also perform saliency analysis [18] to interpret how atom types and surfactant substructures contribute to CMCs and compare these results with the physical rules that correlate structural information of surfactants to CMCs [6]. Following such rules, we propose a small set of new surfactants for which experimental CMCs are not available; for these molecules, we show that CMCs predicted with our GCN exhibited similar trends to those obtained from molecular simulations [19]. The findings of this research provide evidence that GCNs can be used for high-throughput screening of surfactants with desired self-assembly characteristics.

References:

[1] D. Myers, Surfactant Science and Technology: Third Edition, John Wiley and Sons, 2005.
[2] A. Barati, A. Najafi, A. Daryasafar, P. Nadali, H. Moslehi, Adsorption of a new nonionic surfactant on carbonate minerals in enhanced oil recovery: Experimental and modeling study, Chem. Eng. Res. Des. 105 (2016) 55–63.
[3] G.P. Kumar, P. Rajeshwarrao, Nonionic surfactant vesicular systems for effective drug delivery—an overview, Acta Pharm. Sin. B. 1 (2011) 208–219.

[4] F. Gallou, N.A. Isley, A. Ganic, U. Onken, M. Parmentier, Surfactant technology applied toward an active pharmaceutical ingredient: more than a simple green chemistry advance, Green Chem. 18 (2015) 14–19.
[5] M.J. Rosen, J.T. Kunjappu, Surfactants and Interfacial Phenomena: Fourth Edition, John Wiley and Sons, 2012.
[6] J. Israelachvili, Intermolecular and Surface Forces, Elsevier Inc., 2011.
[7] V.P. Torchilin, Structure and design of polymeric surfactant-based drug delivery systems, J. Control. Release. 73 (2001) 137–172.
[8] K.C. Cheng, Z.S. Khoo, N.W. Lo, W.J. Tan, N.G. Chemmangattuvalappil, Design and performance optimisation of detergent product containing binary mixture of anionic-nonionic surfactants, Heliyon. 6 (2020) e03861.
[9] N. Scholz, T. Behnke, U. Resch-Genger, Determination of the Critical Micelle Concentration of Neutral and Ionic Surfactants with Fluorometry, Conductometry, and Surface Tension—A Method Comparison, J. Fluoresc. 28 (2018) 465–476.
[10] C.G. Gahan, S.J. Patel, M.E. Boursier, K.E. Nyffeler, J. Jennings, N.L. Abbott, H.E. Blackwell, R.C. Van Lehn, D.M. Lynn, Bacterial Quorum Sensing Signals Self-Assemble in Aqueous Media to Form Micelles and Vesicles: An Integrated Experimental and Molecular Dynamics Study, J. Phys. Chem. 2020 (2020) 3628.
[11] A. Vishnyakov, M.T. Lee, A. V. Neimark, Prediction of the critical micelle concentration of nonionic surfactants by dissipative particle dynamics simulations, J. Phys. Chem. Lett. 4 (2013) 797–802.
[12] A.P. Santos, A.Z. Panagiotopoulos, Determination of the critical micelle concentration in simulations of surfactant systems, J. Chem. Phys. 144 (2016) 044709.
[13] K. Roy, H. Kabir, QSPR with extended topochemical atom (ETA) indices: Modeling of critical micelle concentration of non-ionic surfactants, Chem. Eng. Sci. (2012).
[14] T. Gaudin, P. Rotureau, I. Pezron, G. Fayet, New QSPR models to predict the critical micelle concentration of sugar-based surfactants, Ind. Eng. Chem. Res. (2016).
[15] X. Li, G. Zhang, J. Dong, X. Zhou, X. Yan, M. Luo, Estimation of critical micelle concentration of anionic surfactants with QSPR approach, J. Mol. Struct. THEOCHEM. 710 (2004) 119–126.
[16] P.D.T. Huibers, V.S. Lobanov, A.R. Katritzky, D.O. Shah, M. Karelson, Prediction of critical micelle concentration using a quantitative structure-property relationship approach. 2. Anionic surfactants, J. Colloid Interface Sci. 187 (1997) 113–120.
[17] Z. Wu, S. Pan, F. Chen, G. Long, C. Zhang, P.S. Yu, A Comprehensive Survey on Graph Neural Networks, ArXiv. (2019).
[18] J. Adebayo, J. Gilmer, M. Muelly, I. Goodfellow, M. Hardt, B. Kim, Sanity Checks for Saliency Maps, Adv. Neural Inf. Process. Syst. 2018-December (2018) 9505–9515.
[19] T. Jin, S.J. Patel, R.C. Van Lehn, Molecular simulations of lipid membrane partitioning and translocation by bacterial quorum sensing modulators, PLoS One. 16 (2021) e0246187.

624289 Cript: Establishing and Harnessing a Polymer Database

Abstract:

Having accessible well-structured data is the foundation of cheminformatics. The complexity of polymer structures poses significant challenges in the formation of databases as there is no single representation that can capture the full molecular detail of a polymer material. More specifically, polymers are large stochastic molecules with distributions in chain length, composition, and topology. Additionally, data collection methods are highly variable, typically provide relative structural information (ex. molecular weight relative to a polystyrene standard), and/or use models which require expert knowledge to put into context. In some cases, experimentally obtaining structural information is impossible, and information from prior processing steps is needed. To complicate matters further, polymers can assemble into a wide range of morphologies through phenomena like phase segregation and crystallization. The formation of these morphologies can be highly influenced by the processing condition under which the material was made. Ultimately, data sets that do not completely capture all the relevant polymer, process, and characterization information pose challenges for advancing data-driven research in the polymer field.

Here, we highlight the efforts of the CRIPT initiative to develop a digital ecosystem for polymer data and the underlying innovation that is making it possible. The CRIPT ecosystem consists of a universal polymer data model, open-source community driven database, machine learning tools, and supporting software. The universal data model provides a unique graph-based representation that captures the full scope of polymer data, which includes synthesis, processing, chemical characterization, physical characterization, and simulation data in a single uniform structure. To support the data model, an open-source online app adhering to FAIR data principles (findable, accessible, interoperable, and reusable) is under development which will serve as a community resource for polymeric material design by allowing researchers to both: add their own data sets and retrieve any other data on the database. To facilitate the population of the database, NLP (natural language processing) techniques are being developed to extract data from prior literature, which will complement the addition of new polymer data by active users. The CRIPT database hopes to provide polymer researchers with access to 'big data' to increase the community's ability to develop better materials faster.

622210 Predicting Adhesive Free Energies of Polymer-Surface Interactions with Machine Learning

Abstract:

Polymer-surface interactions are crucial to biological processes and industrial applications of polymers. A polymer's composition significantly influences its structural and functional properties, such as its solubility and conformations. Here we propose a machine-learning method to connect a model polymer's composition with its adhesion to decorated surfaces. We simulate the adhesive free energies of 20000 unique coarse-grained 1D sequential polymers interacting with functionalized surfaces and build support vector regression models that demonstrate inexpensive and reliable prediction of the adhesive free energy as a function of the sequence. Our work highlights the promising integration of coarse-grained simulation with data-driven machine learning methods for the design of new functional polymers and represents an important step toward linking polymer composition with its derived properties.

626714 Virtual High-Throughput Screening of Vapor-Deposited Amphiphilic Polymers for Biofilm Reduction with Machine/Deep Learning

Abstract:

Researchers identified amphiphilic copolymers as a promising material against biofilm formation on separation membrane surfaces in the recent decade. However, a guide for developing amphiphilic copolymers remains unclear in many aspects (e.g., optimal pairs and compositions of hydrophobic and hydrophilic moieties). Conventionally, one would try to screen material candidates from many experiments and discover a material. This method indeed would provide a path to reach the goal, but it requires a much longer time to explore and improve a testing sample along with costly chemicals and equipment.

For solving these issues, artificial intelligence (AI) could potentially assist or even guide material development while making this discovery more efficient but less expensive. This abstract presents a robust machine/deep learning model across seven data sources and a potential path to fast screen material candidates with AI, which have not been generalized by other computational models for amphiphilic copolymer high-throughput screening.

In our approach with RDKit, our team constructed a synthetic feature library based on the experimental antifouling and antibiofilm performances (quantified by logarithmic fluorescence intensities) of 2,435 unique mixed copolymers against *Pseudomonas aeruginosa* PA01 (PA) suspensions. Here, we labeled each copolymer's quantified performance as Log FPA. After data cleaning, we randomly split the database with stratification (an 80:20 train-test-split) so that both training and test sets followed a similar distribution, and we normalized the training set. Then, we used the same method earlier to split the normalized training set into a sub-training set and a validation set. We built an autoencoder for dimensionality reduction with these two subsets, which kept as much information in a lower dimension and significantly reduced the datasets' noise.

Furthermore, we trained a radial-basis-function-kernelized support vector regressor (SVR) in 5-fold cross-validation (CV) with benchmarks of root mean squared error (RMSE) and coefficient of determination (R2). Meanwhile, for nine different compositions (e.g., from 10:90 to 90:10 with an increment of 10 on the numerator) of amphiphilic copolymers, we developed nine unseen amphiphilic copolymer databases with 61 hydrophobic and five hydrophilic moieties. These two groups of moieties were labeled based on their Log P values (e.g., hydrophobic moiety has a Log P greater than 2, and hydrophilic moiety has a Log P less than 0). Also, we built two databases that solely had either hydrophobic polymers or hydrophilic polymers. Upon passing the final model evaluation, we used this trained model to screen these nine unseen databases and two polymer databases and identified promising materials for experimental validation.

By designing the architecture of the encoding and decoding layers, the autoencoder achieved a reconstruction error of 3.50 * 10e-4 for the sub-training set and 4.05 * 10e-4 for the validation set during the dimensionality reduction step. By applying the 5-fold CV, the SVR model gained the sub-training and validation sets' benchmarks: for a Log FPA range from 5 to 8, the average RMSE was 0.363 (+/- 0.007) and 0.395 (+/- 0.028). The average R2 was 0.642 (+/- 0.011) and 0.576 (+/- 0.045). Then, the trained model achieved an RMSE of 0.360 for the training set and 0.395 for the test set regarding the final model evaluation step. Also, it had an R2 of 0.649 for the training set and an R2 of 0.571 for the test set. In Figure 1 attached in the image file, we present the visualized final model evaluation results. Finally, the model's overall performance for the entire database was as follows: an RMSE of 0.354 and an R2 of 0.660.

For the 11 databases, the trained model screened 2,745 unseen amphiphilic copolymers, 61 hydrophobic polymers, and five hydrophilic polymers. Based on the ranked performances and the vapor-deposition experimental requirements (e.g., volatility of the monomers), our team selected Hexafluoroisopropyl Methacrylate (HFiPMA) and 2-Hydroxyethyl Methacrylate (HEMA) as the amphiphilic copolymer at a composition of 40:60 for experimental validation. We chose this composition because this copolymer under this composition had the optimal predicted Log FPA. Figure 2 attached in the image file presents the predicted Log FPA for this copolymer against all screened compositions.

Our results present a robust model across multiple data sources and a promising material, demonstrating a material screening procedure and guiding the novel material discovery. Beyond this, our team is looking forward to decoding what the model learned and applying a less-data-dependent method to boost our material discovery, such as reinforcement learning.

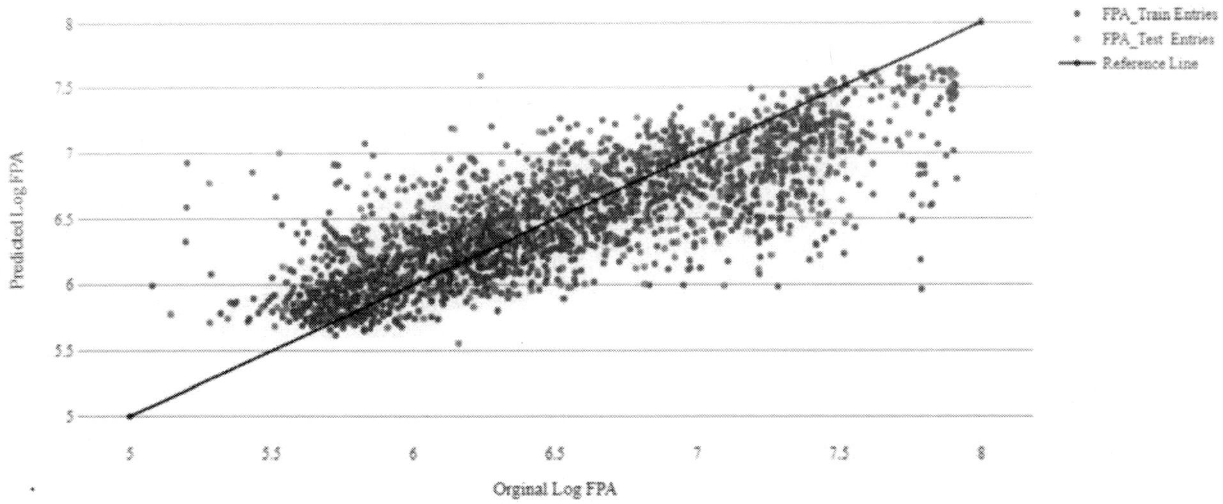

Figure 1. Visualized Final Model Evaluation Results for Log FPA

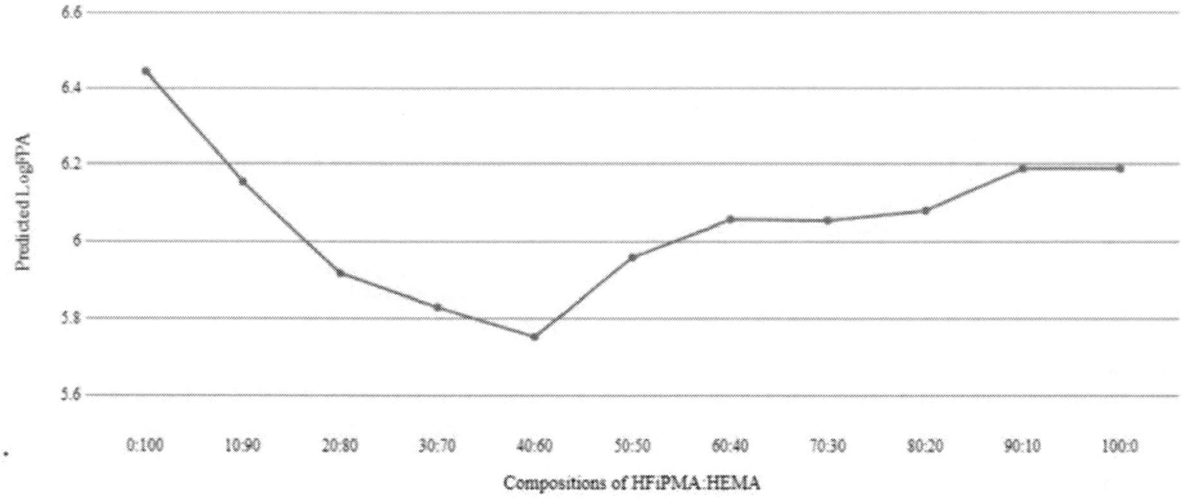

Figure2. Predicted Log FPA for each composition of HFiPMA and HEMA

628736 High-Throughput and Data-Driven Strategies for the Design of Deep Eutectic Solvent Electrolytes.

Abstract:

Deep eutectic solvents (DES) are composed of two or more materials that when mixed possess a lower melting point than either of the individual components. In recent years, both DES and their Ionic Liquid (IL) counterparts have been proposed as electrolytes for various energy storage applications due to their adequate ionic conductivities, particularly for Redox-Flow Batteries (RFBs). A strong emphasis has been placed on RFB materials that are low-cost and environmentally friendly. In this regard, DES have an advantage over ILs because they are composed of cheap, widely available materials with an inexpensive synthesis method. In addition, many of these materials are biodegradable and non-hazardous. However, a major challenge in finding optimum candidates is the overwhelming design space for these materials. The number of possible DES combinations rivals or even exceeds the number of estimated 10^{18} different possible ILs. Current experimental efforts and methods have been slow to make a significant dent in the design space. While computational methods have become a powerful tool to provide further insight where experiments may fall short, a common theme amongst the literature is the lack of data available to make significant strides in developing generalized predictive models for DES. The development of ideal DES electrolytes will require optimization and functional design strategies, along with new methods that can tackle the vast chemical space.

Here, we present our approach to the design of DES electrolytes using high-throughput experimentation and data driven strategies. DES are synthesized by first dissolving the starting components in concentrated stock solutions using volatile solvents. An open-source, automated liquid handling robot then prepares solutions of the components at various molar compositions in multi-well plates. The samples undergo a series of evaporation steps to remove the volatile solvent followed by a final step under vacuum. Liquid samples at room temperature are identified, and the eutectic composition is determined using an in-house constructed and open-source thermal infra-red imaging system able to determine melting points in high-throughput at a fraction of the time compared to traditional methods. Electrochemical characterization to determine conductivities and potential windows is performed in a high-throughput manner using multi-well plates with screen-printed electrodes connected to a standard potentiostat. Finally, approaches to using data science, cheminformatics and engineering metrics to develop design of experiments is presented as a strategy to further probe the chemical space for these materials.

632420 AI-Guided Autonomous Flow Reactor Platform for Accelerated Nanomaterial Synthesis Screening and Parameter Space Mapping

Abstract:

The ability to precisely tune the properties colloidal quantum dots by controlling their composition, structure, size, and size distribution makes them a promising class of materials for applications including ranging from optoelectronics and photovoltaics to bioimaging. Given the vastness of synthesis parameter space, identifying optimal compositions and structures as well associated optimal synthesis recipes for these complex nanocrystals, however, remains a major hurdle for accelerated screening and discovery. The implementation of modular flow reactors, integrated with real-time reaction monitoring, automation and data science, for resource-efficient synthesis has the potential to accelerate this material discovery and synthesis screening process.

Our work utilizes a modular continuous flow reactor platform for synthesis of heavy-metal-free colloidal quantum dots with improved size-uniformity and enhanced optical properties. Our efforts, in collaboration with experts from industry and academia, have led to multiple promising insights into the synthesis of InP-based quantum dots.

The main feature of this presentation will be the development of a flow reactor platform that integrates inline spectroscopy with artificial intelligence-based feedback algorithms for efficient mapping and exploration of the full multidimensional chemical synthesis parameter space (rather than merely identifying a set of synthetic parameters to synthesize a quantum dot with a specific desired properties). This platform autonomously learns the parameter space using 44 experiments, develops a predictive model, and synthesizes InP quantum dots of targeted band gap and polydispersity all through self-driven experiments with no prior knowledge of the reaction space within 28 hours. Our results underscore the promise and critical role of data-science-assisted experimentation for not only accelerating the screening and discovery of colloidal nanocrystals but also towards maximizing synthesis insights across multidimensional parameter space for different classes of colloidal materials.

631399 Machine Learning Models for Rapid Compositional Quantification of Complex Multicomponent Mixtures Using Vibrational Spectroscopy

Abstract:

In recent years, the application of Machine Learning (ML) in chemical sciences has emerged as an attractive method for discovering new materials and reaction mechanisms, optimizing chemical reactors, and automating processes. In this work, we present the development of an analytical methodology based on multitarget regression machine learning models to determine the composition of multicomponent chemical mixtures via Fourier Transform Infrared (FTIR) absorption spectroscopy.

We use principal component regression (PCR) and Artificial Neural Network (ANN) algorithms to train and test models capable of determining the composition of aqueous solutions of unknown concentration. In our approach, we begin by systematically collecting infrared absorption data for training the algorithms. This training data is collected automatically with a set of programmable syringe pumps that prepare and deliver up to ternary aqueous solutions of known concentrations into an FTIR transmission cell where a spectrum from each sample is collected. We then evaluate the ability of commonly used machine learning models trained with this data to predict the concentrations of unknown mixtures. Specifically, machine learning models from publicly available libraries were trained using this data and evaluated for 1-, 2- and 3-components in an aqueous solution. As model complex systems, we used mixtures of components with similar chemical structures such as alcohols (i.e., glycerol, isopropanol, and 1-butanol) and nitriles (i.e., acrylonitrile (AN), adiponitrile (ADN), and propionitrile (PN)). These mixtures are relevant to emerging electrochemical synthesis processes based on glycerol electroreduction and ADN electrosynthesis. They are also challenging because they have similar but subtly different vibrational features, i.e., similar fingerprints. For PCR models, the coefficient of determination, R^2, was 0.982 for the 3-component alcohol mixture and 0.976 for the 3 component nitrile mixture. For ANN models, the accuracy was slightly lower, with R^2 equaling 0.978 and 0.936, respectively. These R^2 correspond to mean absolute errors on the order of 0.07 – 0.24 % m/m for mixtures with component concentrations between 4-10 % m/m. These results suggest that the commonly used machine learning models are capable of and appropriate for determining the unknown composition of complex multicomponent mixtures with similar absorption features. They can potentially be implemented as rapid, in-line, non-invasive, chemical quantification tools for chemical reactor outflow streams analysis.

628555 Are Structure-Function Relationships Governing Polymeric Gene Delivery Payload-Specific?

Abstract:

Polymeric carriers have proven to be safe, economical, and versatile platforms for the intracellular delivery of therapeutic nucleic acid payloads such as DNA, RNA, and genome editing payloads (e.g., ribonucleoproteins or RNP). Clustered regularly interspaced short palindromic repeats (CRISPR) technology requires polymeric carriers to package, protect, and deliver diverse therapeutic payloads, ranging from plasmids (pDNA), RNA, and ribonucleoproteins. Whether structure-function trends identified for a specific nucleic acid payload are broadly applicable to cargoes with contrasting molecular characteristics, sizes, or biological functionalities remains unresolved. Screening a combinatorially designed polymer library, we discovered that P38, which had earlier been identified as the lead polymer for delivering RNPs, proved to be the most promising carrier for pDNA as well. In addition to mediating high levels of transgene expression, P38 effected homology-directed repair (HDR) at higher rates than the commercial polymeric reagent JetPEI while co-delivering RNPs and pDNA donors to HEK293 cells. Further, the superlative pDNA delivery performance of P38 was generalizable to two therapeutically relevant cell types, human mesenchymal stem cells and retinal pigment epithelial cells, thereby validating the broad applicability of P38 across diverse clinical contexts. The versatility of P38 in delivering multimodal cargoes initially suggested that design specifications for RNP and pDNA carriers overlap considerably. However, in contrast to intracellular RNP delivery, polymeric delivery of pDNA is far less reliant on hydrophobic-hydrophilic phase balance and instead exploits electrostatic interactions to realize functional pDNA delivery. By elucidating the physicochemical basis for polyplex internalization, toxicity, and transfection efficiency through statistical learning, we derived payload-specific guidelines that inform the design of bespoke polymers for specific therapeutic contexts.

632081 Enhancing Insight to Individual and Population-Based Microglial Reactivity with Image Analysis

Abstract:

Background:

Organotypic whole hemisphere (OWH) brain slices serve as a high-throughput intermediate between cell cultures and *in vivo* studies that still maintain 3D complexity, regional representation, and the multicellularity of the brain microenvironment [1]. To model ischemic pathology, we developed an oxygen-glucose deprivation (OGD) OWH brain slices model [2] that has demonstrated impact on neurons and microglia. Microglia are the resident immune cells of the brain and can become activated and drive neuroinflammation and ongoing injury after an injurious event. Therefore, microglia have become of interest for therapeutic targets. However, microglia are dynamic after injury and, when analyzed on a population-level, exhibit heterogenous phenotypes that complicate therapeutic strategies. For example, the ischemic hippocampus may exhibit five different phenotypes of microglia, including a resting phenotype not typically associated with injury or injury-driving microglia. We sought to provide a quantitative imaging-based assessment to categorize individual microglial features based on microglia shape features, including perimeter, area, convex area, and circularity, among others, while also analyzing larger populations of microglia to quantify phenotype heterogeneity across treatments, regions, and sex. To quantify entire populations of microglia, we combine our feature analysis with a principal component analysis (PCA) on coordinates positioned along the outside of the segmented cells [3]. The addition of PCA allows us to classify microglia into a number of groups with distinct phenotype representations. Combining the representative phenotypes analyzed on a population level with individual cell features obtained via non-destructive image processing techniques allows us *in situ* insight into comprehensive microglial morphology in the presence of injury.

Methods:

We developed a high-throughput cell quantification pipeline called Frameworks for Fluorescent neuroImage Based Experimental Routines (FFIBER). FFIBER is a seven-step process: develop a data awareness, design a data management plan, determine an optimal experimental pipeline, build out supporting data science infrastructure, perform primary and supplemental imaging, produce interpretable visualizations of results, and build educational modules for developed protocols. For each brain slice dataset, we use FFIBER to develop a routine to quantify cell morphology. For the data science infrastructure, we use Sci-Kit Image in Python to apply an Otsu threshold to segment and label our cells. We then apply geometric algorithms from Sci-Kit image to obtain more than ten morphological features. We manage these features using essential database management in Python. We then split our images into testing, training, and validation sets to feed into the Visually Aided Morphometric Image Recognition (VAMPIRE) package [3]. This package applies PCA to all of our cells and produces a dendrogram of representative cell morphologies. Dendrogram testing data is then fed into the algorithm labeled for key experimental characteristics – sex, treatment, brain region, and species. We then use data visualization concepts to produce interpretable results by generating heat maps (1) of all features by treatment group, and (2) by representative morphology from VAMPIRE. The heat maps of all features are then connected to representative morphology in order to determine correlation of individual cell features such as perimeter, area, convex area, and circularity with representative phenotype distribution across treatment-based, sex-based, and regional microglial populations.

Results:

We applied FFIBER to experimental OWH OGD brain slice data obtained from the rat and the ferret. In the rat OGD brain slice study, we examined microglia as a population and individually. Our population analysis with VAMPIRE showed more heterogeneity in microglia phenotypes after OGD compared to the non-treated control. While, OGD followed by treatment with an anti-inflammatory drug, azithromycin (AZ, an anti-inflammatory drug), had similar heterogeneity of microglia phenotypes compared to the non-treated control. The heterogeneity results are interesting because we observed microglial behavior population-wise and *in situ*. Across all treatment groups, frequency ranking – a ranking of how often a specific phenotype occurs - of each microglial phenotype remains almost exactly the same. However, the frequency difference between each rank is much smaller in the injured group than in the non-treated control and AZ group.

Additionally, we applied geometric analysis to individual microglia to analyze circularity. We observed that the microglia that underwent OGD without treatment were more circular than the microglia in the non-treated control and in the OGD with anti-inflammatory treatment. Microglia with high circularity are correlated with an ameboid phenotype and are associated with driving pathogenic inflammation in the brain. These results supported cytotoxicity data in the rat studies, which signified that OGD results in cell death, including microglia, and can be recovered when treating with AZ. We identified population-wide shifts in microglia phenotypes in the presence of injury that are reversed following treatment with AZ and observed that some of this shift is due to individual microglia taking on a more circular and less ramified structure.

We also applied FFIBER to OGD brain slice data from the ferret. We began by analyzing the representative morphologies of whole microglia populations. We visualized population-wide sex differences in microglia phenotype distributions, including higher reactivity in microglial shape shifts for the male ferret when compared to the female. The larger shift in microglia population phenotypes in the male could be correlated to the clinical differences in sex-based response to injury and treatment in neonatal brain injury. We also delineated and visualized the microglia populations by brain region. When populations were analyzed by region, we observed differences in the reactivity of microglia across the hippocampus, cortex, and corpus callosum – this data aligns with the clinical data showing regional responses to injury in neonates with ischemic brain injury. Finally, we used image processing to quantify circularity, perimeter, area, and convex area of individual cells and were able to correlate the population-based representative phenotypes that have the greatest frequency shifts between non-treated, injured, and treated groups with ramified and bushy microglia phenotypes. Changes in microglia ramification *in situ* correlate with microglia activation in presence of injury and possible deactivation in presence of injury with treatment. All results were obtained using an image processing method that does not alter the microglia or their environment during image procurement or measurement, supporting FFIBER as an *in-situ* cell quantification method.

Conclusions:

By developing quantitative cell morphology pipelines for fluorescent images, we can quantify population-level microglia phenotype presentation and individual microglia features for comprehensive understanding of the dynamic behavior of microglia. With an individual and population-based perspective we can measure sex and regional patterns in cell phenotypes in situ without destroying or altering the cells, which is not possible by other methods. These image processing pipelines combined with the high-throughput nature of brain slice models allowed us to study the effects of OGD across both the rat and ferret species. In the rat, we detected population-wide phenotypic changes in microglia in response to injury that were resolved using AZ. Large shifts in population-based phenotypes were correlated with circularity increases after injury that were not observed in the azithromycin group. In the ferret, we connected representative morphologies found using PCA with features of active microglia such as cell ramification. We were also able to use FFIBER and the brain slice model platforms to perform experiments across different sexes, brain regions, and multiple time points of OGD and treatment. In the future, the development of FFIBER will grow our image processing capabilities to include more algorithms for microglia shape feature analysis including fractal analysis for branching and increase robustness of our software so that cells, including other types of brain and non-brain cells, can be analyzed by users with minimal coding experience.

References:

[1] Liao, R., T.R. Wood, and E. Nance, *Superoxide dismutase reduces monosodium glutamate-induced injury in an organotypic whole hemisphere brain slice model of excitotoxicity.* Journal of Biological Engineering, 2020. **14**(1).

[2] Joseph, A., et al., *Nanoparticle-microglial interaction in the ischemic brain is modulated by injury duration and treatment.* Bioengineering & Translational Medicine, 2020. **5**(3).

[3] Phillip, J.M., et al., *A robust unsupervised machine-learning method to quantify the morphological heterogeneity of cells and nuclei.* Nature Protocols, 2021.

632285 High-Affinity Non-Immunoglobin Scaffolds Promoted By Sequence Mining Impactful in High-Value Therapeutics

Abstract:

Proteins are crucial elements in myriad of biological processes, and they have high value in therapeutics and diagnostics. The global protein engineering market has been predicted to reach $3.9 billion by 2024, highlighting the need for grasping new technologies and opportunities in the field. As a result, for joining this rapidly evolving and valuable campaign, decent understanding on the general regulations and limitations of proteins' realm is needed to ultimately boost or modify their functions by appropriate mutation(s). There are 20 natural amino acids which are building blocks of the proteins combining to each other and forming a relatively long string. This represents an astronomical number of possible mutations even for a small protein with 50 amino acids (20^{50} possibilities). The scarcity of functional proteins and the ease with which deleterious mutations are introduced highlights the need for auxiliary tools in current techniques such as rational design and directed evolution. Recently, machine learning (ML) has been strikingly impactful in protein engineering offering unique advantages such as multitasking and parallelism, high adaptation to complex systems, and feature extraction. The intricate sequence-structure-function paradigm in proteins opens the possibility for various directions in protein engineering from protein design to protein structure prediction. Therefore, ML can be applied in various tasks for predicting both structural and functional attributes of proteins (e.g. secondary structure, stability, and catalytic activity prediction). The number of solved structures from protein sequences, albeit rapidly increasing, are not comparable with the number of growing sequences in protein databases. Hence, there is a growing tendency to map the sequence to function, circumventing structural analysis and predictions, via machine learning techniques. Importantly, mapping the sequence to function for a protein of interest requires a supervised learning algorithm (each protein sequence being labeled with its fitness toward the desired functionality). This goal might be feasible after carrying out high-throughput wet-lab experiments and trying to sample the protein fitness landscape by generating as much labeled data as possible. However, a lot of useful information in unlabeled data will be ignored by this route. Despite progress in biomedical and industrial applications using carefully engineered proteins, a deep understanding for the effect of individual amino acids within a protein and the interconnected relationship of amino acids throughout a protein sequence is still missing. Considering each amino acid as a single word in the language of proteins, natural language processing (NLP) is a promising route to capture the nuances (e.g. epistatic relationships and statistical analysis of amino acids). NLP models can then be built based on the publicly available and rapidly growing databases of annotated protein sequences in order to learn the intricacies of functional amino acid sequences in the context of a protein of interest. The provides the opportunity to then build *supervised* models for fitness prediction in accordance with the previously trained NLP model. In this study, we implement an NLP based method for binding affinity prediction of small non-immunoglobulin scaffolds against biomarkers relevant to treating and diagnosing human disorders. This promising approach will give rise to improved prediction and discovery of high-fitness small scaffolds to aid existing therapeutic and diagnostic technologies.

622104 The Euler Characteristic: A General Topological Descriptor for Complex Data

Abstract:

Datasets are mathematical objects (e.g., point clouds, matrices, graphs, images, field/functions) that have shape [1]. Characterizing the shape (geometrical features) of these objects is important but is not always straightforward. Popular tools from statistics, linear algebra, and signal processing (e.g., moments, correlation functions, singular value decomposition, convolutions, Fourier analysis) do not directly characterize geometrical features of data objects; instead, such tools are used to characterize other types of features (e.g., variance and frequency content).

Topology is a branch of mathematics that provides powerful tools to characterize the shape of data objects. One such tool is the so-called Euler characteristic (EC); the EC, originally used for the characterization of polyhedra [2], is now broadly used in scientific areas such as random fields [3, 4, 5], cosmology [6, 7, 8], material science [9, 10, 11, 12], thermodynamics [13, 14, 15], and neuroscience [16, 17]. To the best of our knowledge, the EC has seen limited applications in engineering; moreover, a fact that is often overlooked in the literature is that the EC provides a general descriptor of different types of topological spaces (this enables the characterization of a wide range of data). This generality arises from the fact that (i) one can use transformations to map a data object into another type of object and (ii) the EC has fundamental connections with statistics, field theory, linear algebra, and graph theory.

Topological descriptors such as the EC offer advantages over statistical descriptors [18]. For instance, statistical descriptors such as Moran's I or correlation matrices do not directly capture the spatial structure of the data (thus limiting the ability to characterize geometrical features) [19, 20]. High-order statistical descriptors such as correlation functions are also limited at capturing spatial and morphological features of the data (especially if the data object is irregular) [20]. However, there exists well-developed theory that connects the EC to the geometry of random fields [4, 5]. Such work establishes that the EC encodes the information of simple statistical descriptors such as means and variances and of more complex descriptors (e.g., space-time covariances) [3]. These connections between topology and statistics are powerful and provide a mechanism to understand the emergence of topological features from physical behavior (e.g., diffusion phenomena). The EC also connects with concepts from linear algebra and graph theory; for example, a matrix or an image can be represented as a weighted graph and the geometry of the graph can be quantified using the EC (a graph is a 2D polyhedron).

In this talk, we will focus on the EC and its application as a descriptor that characterizes geometrical features of data. This characterization is accomplished by performing a decomposition of a data object into a set of independent topological bases which is summarized in the form of what is called an EC curve. We briefly discuss the mathematics of the EC and how it can be used to characterize diverse data objects (e.g. graphs and images/fields). We then shift our focus to the application of these concepts to tackle diverse problems arising in science and engineering; in particular, we discuss how the EC can be used in process monitoring by analyzing correlation structures. We also apply the EC in the analysis of both 2D spatial and 3D spatio-temporal fields; these data objects are derived from reaction-diffusion partial differential equations (PDEs), micrographs of liquid crystals, and flow cytometry. We show that the EC effectively reduces complex datasets and that this reduction facilitates tasks such as visualization, regression, classification, and clustering.

[1] Alexander D Smith, Paweł Dłotko, and Victor M Zavala. Topological data analysis: Concepts, computation, and applications in chemical engineering. *Computers & Chemical Engineering*, page 107202, 2020

[2] Leonhard Euler. Elementa doctrinae solidorum. *Novi commentarii academiae scientiarum Petropolitanae*, pages 109–140, 1758.

[3] Robert J Adler. Some new random field tools for spatial analysis. *Stochastic Environmental Research and Risk Assessment*, 22(6):809–822, 2008

[4] Robert J Adler. *The geometry of random fields*. SIAM, 2010.

[5] Robert J Adler and Jonathan E Taylor. *Random fields and geometry*. Springer Science & Business Media, 2009.

[6] Jens Schmalzing and Krzysztof M Gorski. Minkowski functionals used in the morphological ′ analysis of cosmic microwave background anisotropy maps. *Monthly Notices of the Royal Astronomical Society*, 297(2):355–365, 1998.

[7] Pratyush Pranav, Rien Van de Weygaert, Gert Vegter, Bernard JT Jones, Robert J Adler, Job Feldbrugge, Changbom Park, Thomas Buchert, and Michael Kerber. Topology and geometry of Gaussian random fields i: on Betti numbers, Euler characteristic, and Minkowski functionals. *Monthly Notices of the Royal Astronomical Society*, 485(3):4167–4208, 2019.

[8] M Kerscher, K Mecke, J Schmalzing, C Beisbart, Th Buchert, and H Wagner. Morphological fluctuations of large-scale structure: The PSCZ survey. *Astronomy & Astrophysics*, 373(1):1–11, 2001.

[9] Klaus R Mecke and Dietrich Stoyan. *Statistical physics and spatial statistics: the art of analyzing and modeling spatial structures and pattern formation*, volume 554. Springer Science & Business Media, 2000.

[10] Christoph H Arns, Mark A Knackstedt, and KR Mecke. Reconstructing complex materials via effective grain shapes. *Physical Review Letters*, 91(21):215506, 2003.

[11] Ruth Charney and Michael Davis. The Euler characteristic of a nonpositively curved, piecewise Euclidean manifold. *Pacific Journal of Mathematics*, 171(1):117–137, 1995.

[12] Christian Scholz, Frank Wirner, Jan Gotz, Ulrich Rude, Gerd E Schroder-Turk, Klaus Mecke, and Clemens Bechinger. Permeability of porous materials determined from the Euler characteristic. *Physical review letters*, 109(26):264504, 2012.

[13] KR Mecke. Morphological thermodynamics of composite media. *Fluid Phase Equilibria*, 150:591– 598, 1998.

[14] Hamid Hosseinzade Khanamiri, Carl Fredrik Berg, Per Arne Slotte, Steffen Schluter, and Ole ¨ Torsæter. Description of free energy for immiscible two-fluid flow in porous media by integral geometry and thermodynamics. *Water Resources Research*, 54(11):9045–9059, 2018.

[15] Hendrik Hansen-Goos, Roland Roth, Klaus Mecke, and S Dietrich. Solvation of proteins: linking thermodynamics to geometry. *Physical review letters*, 99(12):128101, 2007.

[16] James M Kilner, Stefan J Kiebel, and Karl J Friston. Applications of random field theory to electrophysiology. *Neuroscience letters*, 374(3):174–178, 2005.

[17] Hyekyoung Lee, Moo K Chung, Hyejin Kang, Bung-Nyun Kim, and Dong Soo Lee. Discriminative persistent homology of brain networks. *In 2011 IEEE international symposium on biomedical imaging: from nano to macro*, pages 841–844. IEEE, 2011.

[18] Eitan Richardson and Michael Werman. Efficient classification using the Euler characteristic. *Pattern Recognition Letters*, 49:99–106, 2014.

[19] Svetlana Lazebnik, Cordelia Schmid, and Jean Ponce. Beyond bags of features: Spatial pyramid matching for recognizing natural scene categories. *In 2006 IEEE Computer Society Conference on Computer Vision and Pattern Recognition (CVPR'06)*, volume 2, pages 2169–2178. IEEE, 2006.

[20] Hubert Mantz, Karin Jacobs, and Klaus Mecke. Utilizing Minkowski functionals for image analysis: a marching square algorithm. *Journal of Statistical Mechanics: Theory and Experiment*, 2008(12):P12015, 2008.

623145 Machine Learning + Automated Reasoning for Theory Discovery

Abstract:

Modern machine learning algorithms find patterns in large datasets, enabling automated search through the high-dimensional spaces of chemicals, materials, reactions, processes, and molecular interactions. However, these data-driven approaches are not easily connected to the equation-based theories that scientists have leveraged and built upon over the past centuries.

We propose a computational framework to discover and derive scientific theories, by integrating two typically distinct areas of computer science: machine learning and automated reasoning. Symbolic regression generates equations that empirically match experimental data. Top equations become "hypotheses" or "conjectures" to be proved or disproved by an automated theorem prover, starting from a set of "axioms" describing the environment under study. We demonstrate this framework by rediscovering and deriving Kepler's Third Law from astronomical observations, and by rediscovering and deriving Langmuir adsorption from experimental measurements.

625173 Characterizing Uncertainty and Error in Machine Learning Chemical Property Prediction

Abstract:

Deep neural networks recently have made a tremendous impact in chemical engineering disciplines, where graph-convolutional neural networks can predict molecular properties with higher accuracy than state-of-the-art techniques. One of the main remaining challenges is to better understand and quantify the different sources of uncertainty associated with molecular property prediction. The uncertainty of the predictions of a machine learning model and thus the deviations from the true target values are characterized by uncertainties due to the model (epistemic uncertainty, including model bias, parameter uncertainty and interpolation uncertainty), as well as uncertainties due to the underlying data and its noise (aleatoric uncertainty). The allocation of the error of a model into epistemic and aleatoric contributions is non-trivial, but essential to characterize possibilities for performance improvements. The categorization of sources of uncertainty in a machine learning model is especially difficult for chemical applications, where the vast chemical space and the diverse nature and number of targets enable a multitude of possible sources for prediction errors. The high dimensionality of chemical space furthermore complicates the differentiation between interpolation and extrapolation for a model prediction.

We systematically study the influence of model bias, such as errors due to the model architecture and input representation, model variance, as well as target data noise on the performance of graph-convolutional neural networks on chemical prediction tasks. Through a clever design of molecular prediction tasks for which an exact solution is known and achievable for a graph-convolutional neural network, we are able to add errors to the data and models in a controlled manner and study the effects on model performance. We combine the addition of controlled errors with different uncertainty estimation techniques, changes to model architecture, and changes in the size or makeup of the dataset to demonstrate trends important to users of machine learning for property prediction. We show that under the influence of random noise in the training and test set, the true performance of a model can continue to improve with larger datasets but the apparent performance will approach an asymptote and cease to improve. Further, we demonstrate the utility of using heteroscedastic and homoscedastic loss functions to assess the presence of noise errors in the dataset, when those errors are associated with model features and when they are not. We apply measured ensemble variance as a method of assessing epistemic error and use statistical analysis of the results to project how much of the model error is due to variance that can be observed with ensembling and how much is a baseline bias. Using trends over batch size and observed interactions between different uncertainty characterizations, we provide methods for estimating the contribution of different error types to the model performance, the likely effects of adding more data, and the maximum benefit available from ensembling.

627055 Giving Attention to Generative Models for *De Novo* *M*olecular Design

Abstract:

Attention mechanisms have led to many recent breakthroughs in sequential data modeling tasks including machine translation, text generation and protein structure prediction.[1–3] While they have also been used in the molecular domain for tasks such as graph-based analysis of chemical structures[4] and atom-mapping of organic reactions[5], attention has yet to be incorporated into any generative algorithms for molecular design.

Here we explore the impact of adding self-attention layers to generative β-VAE models and show that those with attention are able to learn a complex "molecular grammar" while improving performance on downstream tasks such as accurately sampling from the latent space ("model memory") or exploring novel chemistries not present in the training data. There is a notable relationship between a model's architecture, the structure of its latent memory and its performance during inference. For instance, we find that there is an unavoidable tradeoff between model exploration and validity that is a function of the complexity of the latent memory. However, novel sampling schemes may be used that optimize this tradeoff.

We also demonstrate the ability of the transformer VAE to construct a set of complex, human-interpretable molecular substructural features in an unsupervised fashion. We compare these learned features across different input representations including SMILES and SELFIES[6] strings as well as those extracted from traditional cheminformatics software packages. Finally, we discuss how these models may eventually be used in tandem with natural language models, high-throughput molecular dynamics simulations and reinforcement learning algorithms to present a unified AI-based framework for molecular discovery and optimization.

1. Bahdanau, D., Cho, K. & Bengio, Y. Neural Machine Translation by Jointly Learning to Align and Translate. *arXiv:1409.0473 [cs.CL]* (2014).
2. Brown, T. B. *et al.* Language Models are Few-Shot Learners. in *34th Conference on Neural Information Processing Systems (NeurIPS 2020), Vancouver, Canada* (arXiv, 2020).
3. Service, R. F. 'The game has changed.' AI triumphs at protein folding. *Science* **370**, 1144–1145 (2020).
4. Payne, J., Srouji, M., Yap, D. A. & Kosaraju, V. BERT Learns (and Teaches) Chemistry. *arXiv:2007.16012 [q-bio.BM]* (2020).
5. Schwaller, P., Hoover, B., Reymond, J.-L., Strobelt, H. & Laino, T. Extraction of organic chemistry grammar from unsupervised learning of chemical reactions. *Science Advances* **7**, eabe4166 (2021).
6. Krenn, M., Häse, F., AkshatKumar, N., Friederich, P. & Aspuru-Guzik, A. Self-referencing embedded strings (SELFIES): A 100% robust molecular string representation. *Machine Learning: Science and Technology* **1**, (2020).

627282 Improving Data Sub-Selection for Supervised Tasks with Principal Covariates Regression

Abstract:

Data analyses based on linear methods constitute the simplest, most robust, and transparent approaches to the automatic processing of large amounts of data for building supervised or unsupervised machine learning models. Principal covariates regression (PCovR) is an underappreciated method that interpolates between principal component analysis and linear regression and can be used to conveniently reveal structure-property relations in terms of simple-to-interpret, low-dimensional maps. We have recently introduced methods that incorporate PCovR into two popular data selection approaches, CUR and Farthest Point Sampling, which iteratively identify the most diverse samples and discriminating features. While our approach is completely general, here we focus on systems relevant to atomistic simulations, chemistry, and materials science -- fields where feature and sample selection are an increasingly common practice. Our results show that these selection methods identify data subsets that out-perform their unsupervised counterparts--which we demonstrate with models of increasing complexity, from ridge regression to kernel ridge regression and finally feed-forward neural networks.

This work pulls from:

Structure-Property Maps with Kernel Principal Covariates Regression

BA Helfrecht, RK Cersonsky, G Fraux, M Ceriotti

Machine Learning: Science and Technology 1 (4)

Improving Sample and Feature Selection with Principal Covariates Regression

RK Cersonsky, BA Helfrecht, EA Engel, M Ceriotti

arXiv preprint arXiv:2012.12253

631972 Chemical Identification in Multicomponent Electrolytes Using Voltammetry, Physics-Based Modeling, and Bayesian Inference

Abstract:

The prospects of a renewably powered grid continue to grow with the decreasing cost of sustainable energy technologies [1,2]. However, the intermittency of variable resources inhibits their deployment, necessitating energy management systems capable of operating across a range of time scales, such as rechargeable batteries (e.g., lithium-ion batteries, redox flow batteries). These batteries are complex devices whose performance and durability are dictated by the interplay of many constituent components (e.g., electrodes, membranes/separators), with the liquid electrolyte being especially important for long-term operation. Understanding and mitigating electrolyte decomposition within electrochemical systems is key to extending lifetime [3-5], and studies that probe electrolyte degradation are typically conducted using *post mortem* analysis to retroactively elucidate the decay pathway(s) [4-6]. While this approach is largely successful, it relies on a suite of *ex situ* techniques (e.g., nuclear magnetic resonance, mass spectrometry, and UV-Vis spectroscopy) that are time-consuming and expensive. Further, preparatory steps necessary for these *ex situ* analyses (e.g., dilution) can alter electrolyte composition, potentially providing incomplete information about electrolyte behavior. To this end, machine learning algorithms are emerging that label electroactive compounds from *in situ* electroanalytical techniques such as voltammetry [7,8]. While noteworthy, existing protocols can be improved; some methods identify compounds less accurately when the training and testing data are obtained under different electrolyte or experimental conditions [7], while others do not evaluate all possible species combinations in multicomponent solutions [8].

Building upon this prior work, we have developed a versatile protocol that uses physical modeling and binary hypothesis testing to identify redox-active compounds in multicomponent electrolytes. The protocol references a compound library that catalogues physical descriptors (e.g., redox potential, diffusion coefficients) and compares the library to experimental data using Bayesian inference to identify the redox-active species present. The compound identities are then reported to the user, the knowledge of which can be used to perform more targeted *ex situ* experiments by narrowing the feasible set of candidate compounds. This process, in turn, can simplify and streamline electrolyte analysis and may consequently accelerate the development of electrochemical technologies. In this presentation, we will describe protocol development and validation using a model set of phenothiazine derivatives across two distinct voltammetric techniques. We will also discuss how this approach may be used to monitor the complex phenomena underlying electrolyte behavior in practical electrochemical systems.

Acknowledgments: This work was supported as part of the National Science Foundation (NSF) under Award Number 1805566. Any opinion, findings, and conclusions or recommendations expressed in this material are those of the authors and do not necessarily reflect the views of the NSF. We also gratefully acknowledge the MIT Supercloud and Lincoln Laboratory Supercomputing Center for providing HPC resources that have contributed to the research results reported within this work. We finally thank Professor Susan Odom, Dr. Aman Kaur, and the Odom Research Group at the University of Kentucky for synthesizing, purifying, and shipping phenothiazines.

References:

(1) Bolinger, M.; Seel, J. Utility-Scale Solar: Empirical Trends in Project Technology, Cost, Performance, and PPA Pricing in the United States - 2018 Edition; 2018.

(2) Lazard. Lazard's Levelised Cost of Energy Analysis - Version 13.0; 2019; Vol. 13.

(3) Day, R. P.; Xia, J.; Petibon, R.; Rucska, J.; Wang, H.; Wright, A. T. B.; Dahn, J. R. Differential

Thermal Analysis of Li-Ion Cells as an Effective Probe of Liquid Electrolyte Evolution during Aging. Journal of The Electrochemical Society **2015**, 162 (14), A2577–A2581. https://doi.org/10.1149/2.0181514jes.

(4) Goulet, M.-A.; Aziz, M. J. Flow Battery Molecular Reactant Stability Determined by Symmetric Cell Cycling Methods. Journal of The Electrochemical Society **2018**, 165 (7), A1466–A1477. https://doi.org/10.1149/2.0891807jes.

(5) Narayana, K. A.; Casselman, M. D.; Elliott, C. F.; Ergun, S.; Parkin, S. R.; Risko, C.; Odom, S. A. *N*-Substituted Phenothiazine Derivatives: How the Stability of the Neutral and Radical Cation Forms Affects Overcharge Performance in Lithium-Ion Batteries. ChemPhysChem **2015**, 16 (6), 1179–1189. https://doi.org/10.1002/cphc.201402674.

(6) Kwabi, D. G.; Lin, K.; Ji, Y.; Kerr, E. F.; Goulet, M.-A.; De Porcellinis, D.; Tabor, D. P.; Pollack, D. A.; Aspuru-Guzik, A.; Gordon, R. G.; Aziz, M. J. Alkaline Quinone Flow Battery with Long Lifetime at pH 12. Joule **2018**, 2, 1894–1906. https://doi.org/10.1016/j.joule.2018.07.005.

(7) Ye, J.-J.; Lin, C.-H.; Huang, X.-J. Analyzing the anodic stripping square wave voltammetry of heavy metal ions via machine learning: Information beyond a single voltammetric peak. Journal of Electroanalytical Chemistry **2020**, 113934. https://doi.org/10.1016/j.jelechem.2020.113934.

(8) Dean, S. N.; Shriver-Lake, L. C.; Stenger, D. A.; Erickson, J. S.; Golden, J. P.; Trammell, S. A. Machine Learning Techniques for Chemical Identification Using Cyclic Square Wave Voltammetry. Sensors (Switzerland) **2019**, 19, 2392. https://doi.org/10.3390/s19102392.

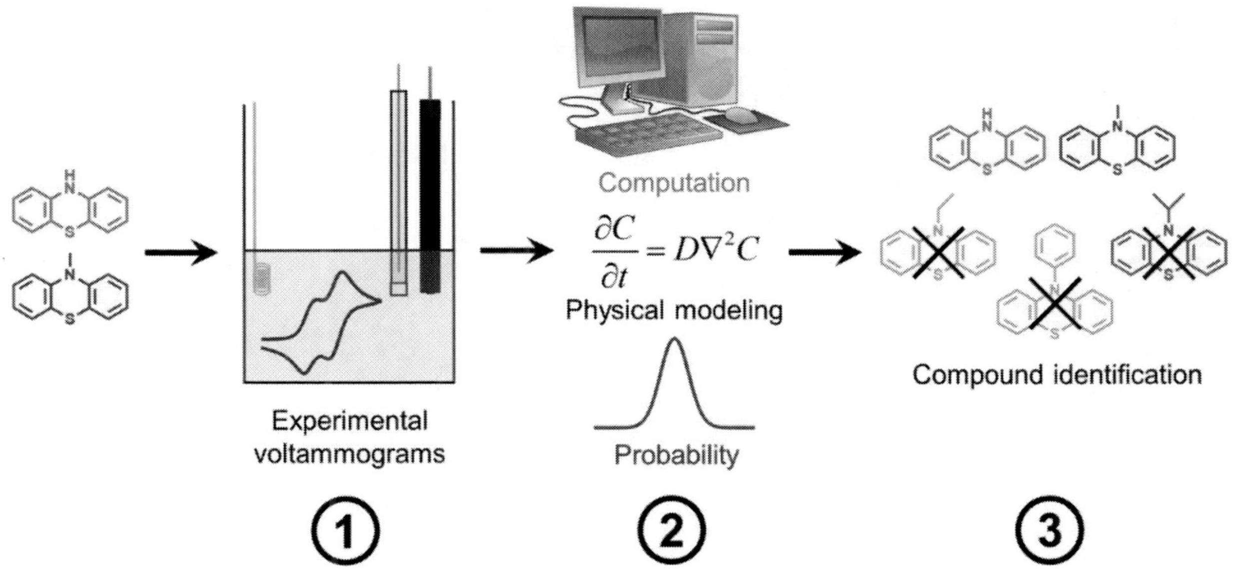

Workflow of the protocol as applied to the phenothiazine derivatives evaluated in this case study. First, an experimental voltammogram of the electrolyte is acquired. Second, a library built using physical modeling is compared to the voltammogram using a computational protocol that uses Bayesian inference. Finally, the protocol reports the compounds present in the electrolyte of interest.

631425 Deep Learning-Assisted Analysis of Anomalous Nanoparticle Surface Diffusion in Liquid Phase Transmission Electron Microscopy

Abstract:

The motion of nanoparticles near surfaces is of fundamental importance in physics, biology, and chemistry. Liquid cell transmission electron microscopy (LCTEM) is a promising technique for studying motion of nanoparticles with high spatial resolution. Yet, the lack of understanding of how the electron beam of the microscope affects the particle motion has held back advancement in using LCTEM for in-situ single nanoparticle and macromolecule tracking at interfaces. Here, we experimentally studied the motion of a model system of gold nanoparticles dispersed in water and moving adjacent to the silicon nitride membrane of a commercial liquid cell in a broad range of electron beam dose rates. We find that the nanoparticles exhibit anomalous diffusive behavior modulated by the electron beam dose rate. We characterized the anomalous diffusion of nanoparticles in LCTEM using a convolutional deep neural-network model inspired by the canonical statistical tests. The results demonstrate that the nanoparticle motion is governed by fractional Brownian motion at low dose rates, resembling diffusion in a viscoelastic medium, and continuous-time random walk at high dose rates, resembling diffusion on an energy landscape with pinning sites.[1]

[1] Jamali, V., Hargus, C., Ben-Moshe, A., Aghazadeh, A., Ha, H. D., Mandadapu, K. K., & Alivisatos, A. P. (2021). Anomalous nanoparticle surface diffusion in LCTEM is revealed by deep learning-assisted analysis. *Proceedings of the National Academy of Sciences*, *118*(10).

622868 A Principal Component Normalizing Flow for Modeling Renewable Electricity Generation

Abstract:

Wind and solar power depend on natural occurrences and are therefore volatile and uncertain. The inherent uncertainty of renewable electricity sources must be considered in decision-making such as design and operation of energy systems, e.g., through stochastic programming (SP) based on discrete scenarios [1-3]. In SP, the scenarios must accurately represent the underlying distributions of the stochastic processes to avoid skewed or infeasible decisions. Traditional scenario generation approaches rely on Gaussian Copulas [3] and autoregressive moving average (ARIMA) models [2-4]. Recent advances in deep learning have enabled scenario generation without prior assumptions using neural autoregressive models (NARX) [5] and deep generative models, e.g., generative adversarial networks (GAN) [6] or normalizing flows [7]. In contrast to NARXs and GANs, normalizing flows model the target distribution explicitly using a nonlinear transformation of a multivariate Gaussian based on invertible neural networks. The inverse of that transformation models the probability density function directly and enables training by log-likelihood maximization [8]. However, recent research from the field of image generation has shown that standard normalizing flows are unable to model manifold distributions and instead assign non-zero probabilities to out-of-distribution data, which results in the generation of noisy data [9]. We find that energy time series often reside on manifolds with a much lower dimensionality than the number of time steps due to the correlation between time steps and that standard normalizing flows should therefore not be used for scenario generation of renewable electricity generation [10]. To maintain the direct likelihood maximization but avoid the generation of noisy data, we propose a dimensionality reducing normalizing flow layer based on the principal component analysis (PCA) [10]. The PCA sets up the normalizing flow in a lower-dimensional space and excludes the space outside of the manifold, i.e., the space of zero probability. We apply the resulting principal component flow (PCF) for scenario generation of solar and wind electricity generation based on historical data from Germany in the years 2013 to 2015. A comparison between standard normalizing flows and the PCF shows a significant improvement in the distribution fit and the representation of the frequency behavior as a result of the dimensionality reduction.

References:

[1] A. Mitsos, N. Asprion, C. A. Floudas, M. Bortz, M. Baldea, D. Bonvin, A. Caspari, P. Schäfer, Challenges in process optimization for new feed-stocks and energy sources, Computers & Chemical Engineering 113 (2018) 209–221

[2] Schäfer, P., Westerholt, H. G., Schweidtmann, A. M., Ilieva, S., & Mitsos, A. (2019). Model-based bidding strategies on the primary balancing market for energy-intense processes. *Computers & Chemical Engineering*, *120*, 4-14.

[3] J. M. Morales, A. J. Conejo, H. Madsen, P. Pinson, M. Zugno, Integrating Renewables in Electricity Markets: Operational Problems, Vol. 205, Springer Science & Business Media, 2013.

[4] Ierapetritou, M. G., Wu, D., Vin, J., Sweeney, P., & Chigirinskiy, M. (2002). Cost minimization in an energy-intensive plant using mathematical programming approaches. *Industrial & engineering chemistry research*, *41*(21), 5262-5277.

[5] S. I. Vagropoulos, E. G. Kardakos, C. K. Simoglou, A. G. Bakirtzis, J. P. Catalao, Ann-based scenario generation methodology for stochastic variables of electric power systems, Electric Power Systems Research 134 (2016) 9–18.

[6] Y. Chen, Y. Wang, D. Kirschen, B. Zhang, Model-free renewable scenario generation using generative adversarial networks, IEEE Transactions on Power Systems 33 (3) (2018) 3265

[7] L. Zhang, B. Zhang, Scenario forecasting of residential load profiles, IEEE Journal on Selected Areas in Communications 38 (1) (2020) 84–9

[8] G. Papamakarios, E. Nalisnick, D. J. Rezende, S. Mohamed, B. Lakshminarayanan, Normalizing flows for probabilistic modeling and inference, arXiv preprint arXiv:1912.02762, 2021

[9] J. Brehmer, K. Cranmer, Flows for simultaneous manifold learning and density estimation, arXiv preprint arXiv:2003.1391, 2020

[10] E. Cramer., A. Mitsos, R. Tempone, and M. Dahmen. "Principal component density estimation for scenario generation using normalizing flows", Manuscript in preparation, 2021.

626059 Lessons and Opportunities in Data-Driven High Throughput Experimentation

Abstract:

The adoption of data-science methods within chemical engineering is primed to revolutionize materials discovery and to accelerate developments in molecular understanding. Yet, the movement is largely dominated by data and information originating from molecular modelling and simulation, or that is acquired from legacy sources (e.g. databases, literature mining) accumulating data from decades of careful experimental work. In order to advance data-driven materials developments well into the future, high-throughput experimentation (HTE) and automation throughout the complete laboratory workflow must also be developed and widely adopted to accelerate rates of experimental data production. In this talk we outline several examples and experiences showcasing how our research group is developing and adapting hardware and software infrastructure to accelerate the pace of molecular discovery in soft-matter systems for applications in health care, clean energy and materials synthesis. The talk will highlight recent research examples related to the implementation of HTE for electrolyte discovery and colloidal formulation/synthesis. We will also highlight significant challenges that emerged while transitioning established 'wet-laboratory' practices to HTE. These relate to adapting routine experimental methods to HTE, developing new skills within the research workforce, the adoption of good data stewardship practices, needs for autonomous data sorting/classification, algorithms for automatic modeling and analysis amongst many others. Conversely, we will also highlight the opportunities that emerge for enhancing virtual collaboration, improving information quality, enabling open data/hardware/software sharing, tackling irreducible problems (e.g. optimization of complex formulations), and outlooks for the implementation of self-driving laboratories.

625523 Automating Discovery: Accelerating the Search for Novel Perovskites with Robots and Data

Abstract:

Scientific discovery has typically been conducted in a bespoke fashion, where human experts design and perform experiments in a manual way. Advances in laboratory automation, data collection, and machine learning may enable new approaches to automating the discovery process.

In this talk, I will describe our Robotic-Accelerated Perovskite Investigation and Discovery (RAPID) project. The first generation of RAPID uses inverse temperature crystallization (ITC) to grow halide perovskite single crystals for x-ray structure determination and bulk characterization using commercial liquid handling robots. The second generation of RAPID uses antisolvent vapor diffusion, expanding the types of chemical processes we can study.

Experiment plans for the syntheses are contributed remotely, by both human scientists and algorithms trained on the reaction data, facilitated by our ESCALATE (Experiment Specification, Capture and Laboratory Automation Technology) data management software. ESCALATE is a general-purpose software system that captures and processes the comprehensive data and metadata collected during experiments into a form suitable for machine learning. I will discuss ESCALATE deployments to problems of halide perovskite crystal growth, perovskite thin film, and solid-liquid and liquid-liquid separations of actinide mixtures. In addition to presenting an abstraction layer for algorithmic-controlled experiments, the comprehensive dataset captured by ESCALATE enables "automated serendipity," wherein we can use statistical analyses to identify hypotheses about how crystal growth depends upon poorly-controlled experimental conditions, and followed by deliberate, automated, interventional studies to confirm or reject those hypotheses.

Throughout the talk I will describe case studies about how we have used this RAPID/ESCALATE to extract new scientific insights, enhance the replicability of our scientific work, and reduce the time to discovery, which is applicable to a variety of discovery projects.

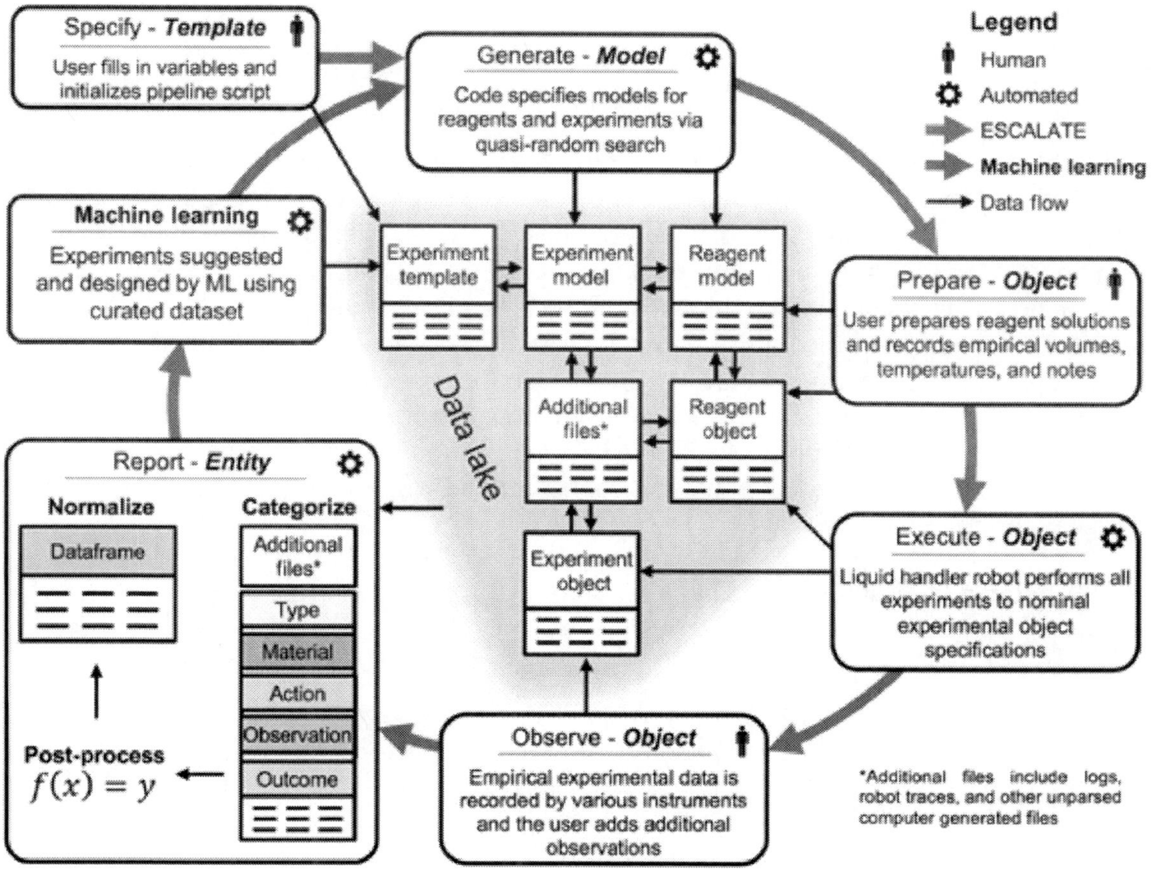
Specify - Template
User fills in variables and initializes pipeline script
Generate - Model
Code specifies models for reagents and experiments via quasi-random search
Legend
Human
Automated
ESCALATE
Machine learning
Data flow
Machine learning
Experiments suggested and designed by ML using curated dataset
Experiment template
Experiment model
Reagent model
Prepare - Object
User prepares reagent solutions and records empirical volumes, temperatures, and notes
Additional files*
Reagent object
Data lake
Report - Entity
Normalize
Categorize
Dataframe
Additional files*
Type
Material
Action
Observation
Outcome
Post-process
f(x) = y
Experiment object
Execute - Object
Liquid handler robot performs all experiments to nominal experimental object specifications
Observe - Object
Empirical experimental data is recorded by various instruments and the user adds additional observations
*Additional files include logs, robot traces, and other unparsed computer generated files

621400 Streamlining Small Molecule Design and Synthesis

Abstract:

This talk will cover progress in the application of data science and machine learning techniques to problems in synthetic chemistry as they relate to molecular discovery. The typical discovery paradigm is an iterative process of designing candidate compounds, synthesizing those compounds, and testing their performance, where each repeat of this cycle can require weeks or months. Time and cost constraints may necessitate selecting compounds on the basis of what are perceived as fast to synthesize, rather than what are most informative to assay. Data science and statistical learning offer unprecedented opportunities to systematize and streamline the process by which new functional small molecules are designed and synthesized. I will describe how historical reaction data can be used to inform decision-making in small molecule pathway design, for both retrosynthesis and forward reaction prediction, and how these techniques can be integrated with *de novo* molecular design to triage candidate compounds for testing.

626319 Building Catalytic Descriptors with Iterative Bayesian Additive Regression Trees (iBART)

Abstract:

Catalyst design can be accelerated by finding physical descriptors that predict the adsorption strength of molecules and atoms on metal and metal-oxide surfaces. These descriptors can be constructed from physical intuition when the systems under investigation are relatively simple, when only one physical phenomenon dominates the interaction between the surface and the adsorbate. However, it becomes increasingly difficult to design such descriptors in complex systems when multiple phenomena influence the interaction between the adsorbate and the surface. Take for example the binding energy of a metal atom and an oxide support, which is a critical metric for controlling the supported metal's cluster-size distribution, sintering rate, and stability. This interaction is influenced by several charge transfers occurring between the metal, the oxide surface, surface defects (e.g., vacancies or heteroatom dopants), and adsorbates from the reaction environment. This complexity cannot be captured by simple physical descriptors derived from chemical intuition alone. This issue can be addressed with statistical learning (SL) techniques that apply feature engineering (FE) and feature selection (FS) to build complex descriptors, as shown in our previous work [1-2]. These studies demonstrate the capabilities of various state-of-the-art methods for FS, yet we found that these methods tend to suffer from two critical deficiencies. First, the FE procedures generate highly correlated feature spaces that challenge weaknesses in typical FS techniques (e.g., LASSO). This leads to higher error and an earlier onset of overfitting. Second, the FS methods require significant computational resources to handle the large features spaces that are necessary for constructing features with sufficient complexity. In this work, we developed a new FE and FS methodology, called iterative Bayesian additive regression trees (iBART), to address these issues. iBART reduces the size of the candidate feature space from millions/billions to hundreds without sacrificing feature complexity, which accelerates the computational speed and lessens the required system memory. We applied iBART to derive metal/oxide binding energy descriptors from our previously published data [1-2], and the results demonstrate that iBART improves descriptor accuracy and significantly reduces computational expense.

[1] Nolan J. O'Connor, A. S. M. Jonayat, Michael J. Janik, and Thomas P. Senftle. "Interaction Trends between Single Metal Atoms and Oxide Supports Identified with Density Functional Theory and Statistical Learning." *Nature Catalysis*, 1 (2018): 531–39.

[2] Chun-Yen Liu, Shijia Zhang, Daniel Martinez, Meng Li, and Thomas P. Senftle. "Using Statistical Learning to Predict Interactions between Single Metal Atoms and Modified MgO(100) Supports." *npj Computational Materials*, 6 (2020): 102.

629694 Unsupervised Machine Learning to Extract the Electronic and Chemical Properties of Alloy and Metal Oxide Surfaces

Abstract:

Catalyst design greatly benefits from the identification of the atomic, chemical, and electronic descriptors that map to catalytic figures of merit (e.g., stability, activity, selectivity). One important and widely used class of descriptors in catalysis is electronic-structure descriptors, which have been extensively used as site-specific chemisorption descriptors for a variety of different materials. Nonetheless, there exists no unified framework to systematically identify electronic-structure descriptors. We demonstrate a simple, yet informative unsupervised machine learning strategy using principal component analysis (PCA) that rapidly identifies accurate and interpretable electronic-structure descriptors of catalysts. We first demonstrate the approach for transition metal alloys and show that the principal component descriptors yield accurate ML models of chemisorption of adsorbates (e.g., O*, C*, H*, N*) that outperform ML models built using traditional electronic structure descriptors. We use an ML model interpretation method called partial dependence plots to visualize the relationships between the identified electronic-structure descriptors and the chemisorption trends. Importantly, we are able to interpret the meaning of these descriptors and connect them to the alloys' geometric structure and composition using signal reconstruction methods from the cognitive neuroscience and machine intelligence literature. Interpretation of the PCs suggests that they capture physical trends that are consistent with prior efforts to develop electronic-structure chemisorption descriptors for alloys. We show that the approach is general by finding descriptors of surface oxygen reactivity for metals and metal oxides. Ultimately, this unsupervised learning approach provides a pathway for finding electronic-structure descriptors for catalytic materials that readily connect to geometric structure and composition.

637980 Implementing and Accelerating Autonomous Closed-Loop Control

Abstract:

Bayesian Optimization is a new machine learning method that can accelerate the path from automation to autonomous manufacturing. We will discuss the benefits of Bayesian Optimization, how it compares to more traditional methods such as APC and MPC, and how it can be used to optimize a real fermentation process. More broadly, we will show how this new technology enables accelerated DoE, product development, and fully autonomous manufacturing with closed-loop AI control.

638841 State of Art in Modeling and Modeling Deployment for the Process and Pharma Industries; from Advanced Simulations to Web Apps

Abstract:

Taking the drug manufacturing process as an example, the presentation will highlight the breadth and value of M&S when it comes to manufacturing. Starting from upstream processes like mixing and clean-in-place, to downstream processes like chromatography, separation, and fill/finish, and ending with packaging and shipping, the value of M&S will be highlighted. Furthermore, the concept of simulation democratization will be introduced where advanced simulations will be made accessible to plant/manufacturing engineers using simple to use Web-Apps."

632267 Analytical Methods to Improve Diagnostic Protocols Using Infrared Spectroscopic Imaging

Abstract:

Spectrometry coupled to data mining approaches can identify spectral signatures indicative of disease state and its progression. Both structural and biochemical changes accompany cancer development and its subsequent progression. Current histologic characterization is morphology-based; thin (5 μm) tissue sections are stained, and cells are visually recognized by a pathologist using an optical microscope. However, the basis of the disease is well known to be molecular. Molecular analysis for pathology is complicated by the spatial diversity of cells and acellular materials, necessitating an analytical technique that involves imaging. New advancements in infrared (IR) spectroscopic imaging techniques have generated big data warranting analytical advances in using spectroscopic imaging for digital histopathology. Biological classes often have subtle variations in the molecular imaging data, genetics, proteomics, or other systems biology experiments. However, new advances in feature extraction and computational analysis have enabled the differentiation of intricate disease categories. In addition to measurement and computational factors, sample size estimation for desired statistical significance and generation of annotated data also influences the development of robust models based on spectroscopic data. Underestimating the sample size can lead to statistically deficient diagnostic tests, for example, while an overestimation can significantly increase experimental costs and time to develop protocols. Additionally, the transfer of annotations from clinical to spectroscopic data can be time-consuming, cumbersome, and have interobserver variability. For accurate classification models downstream, an accurate mapping of ground truth data to the images of interest (IR data) is crucial. Multimodal image registration is typically used to map data from one imaging modality to another. They have been utilized to correlate multiple biomarkers in conventional staining data[1]. Previous studies have also reported the use of image registration approaches for aligning mass spectroscopic images with Raman imaging data[2] and optical images[3]. However, there are still challenges to accurately transfer annotations from stained images (typically used in clinics) to IR data in a user-friendly manner.

In this study, we carry out multivariate analysis of variance (MANOVA) to estimate the discrimination potential of IR spectroscopic features for building accurate machine learning models to separate different diagnostic categories. Next, we build a control point registration-based automated annotation tool that can generate training data for building new models with large-scale validation. The user can precisely annotate coordinates (three points) in the IR image and the clinical image, corresponding to the same spatial architecture using a graphical user interface. This overcomes the limitations of sparse ground truth data with current manual approaches by providing a tool to transfer pathologist annotations from stained images to IR images across diagnostic categories. Spectral features for disease classification are typically selected using feature selection approaches and subsequently used in artificial intelligence (AI) algorithms for diagnosis. Deep learning offers a new approach to combine spatial-spectral features with automatic identification by the AI algorithm. However, deep learning approaches require an extensive database of labeled training data. This would significantly reduce the amount of time needed to generate labeled training data, paving the way for an accurate, fully automated deep learning-based spectroscopic analysis of histopathological samples. We also utilize simple machine learning models to further increase the accuracy of our registration tool. Finally, we develop a combinatorial data mining approach (supervised + unsupervised) to identify diagnostic patterns and selecting pure chemical pixels for each cell type.

References

1. G. Lippolis, A. Edsjö, L. Helczynski, A. Bjartell, N.C. Overgaard, A. Jemal, et al. "Automatic registration of multi-modal microscopy images for integrative analysis of prostate tissue sections". BMC Cancer. 2013. 13(1): 408.
2. T. Bocklitz, K. Bräutigam, A. Urbanek, F. Hoffmann, F. von Eggeling, G. Ernst, et al. "Novel workflow for combining Raman spectroscopy and MALDI-MSI for tissue based studies". Anal. Bioanal. Chem. 2015. 407(26): 7865–7873.
3. T. Gregory Schaaff, J.M. McMahon, P.J. Todd. "Semiautomated analytical image correlation". Anal. Chem. 2002. 74(17): 4361–4369.

631667 Message Passing Neural Networks for Prediction of IR Spectra

Abstract:

Spectroscopy is an important tool for manufacturing for ascertaining purity in biological systems and drug manufacturing. Prediction of unknown spectra would be a powerful tool for extending existing spectral identification databases and evaluating novel molecules for which there exist no available standards. Some recent advances have used machine learning techniques in the prediction of IR spectra from specific molecule classes, such as polyaromatic hydrocarbons. In this work we discuss our release of a model for IR prediction with a more general scope of application. This model will provide researchers with easy access to IR spectra predictions, and the software used to develop the model (an extension of the Chemprop software) will inform efforts to develop other spectra prediction models in the future. The software is freely distributed through GitHub and the trained model files are available through Zenodo.

The presented model for IR spectra prediction requires only the input of a molecular SMILES strings to generate predictions. The chemical representation is processed through a message passing neural network to encode the molecule in a latent vector representation followed by a feed forward neural network for prediction of IR spectra. The entire process is differentiable, making even the encoding of the latent molecule vector learnable and optimizable. The model is pre-trained using semi-empirical quantum chemistry calculations (GFN2-xTB) for molecules sampled from the PubChem database to learn molecular encodings for a wide scope of molecules. Further training is performed using 56,955 experimental spectra collected from four data sources: the National Institute of Standards and Technology (NIST), Pacific Northwest National Labs (PNNL), The National Institute of Advanced Industrial Science and Technology (AIST), and the Coblentz Society. The model allows for predictions of spectra in the gas phase and in four supported condensed phases.

625361 Transfer Learning for Prediction of Absorption and Emission Spectra from Multi-Fidelity Data

Abstract:

Accurate predictions of optical properties from chemical structure are necessary for the design of optimal dye molecules towards important applications such as solar cells, bioimaging, display technologies, etc. The recent compilation of several large datasets of experimental absorption and emission spectral data from the literature has enabled machine learning methods to equal or exceed the prediction accuracy of first principles approaches like time-dependent density functional theory (TD-DFT) at a fraction of the computational cost. Depending on the choice of exchange-correlation functional and the level of approximation, TD-DFT may show systematic and random errors with respect to the experimental ground truth. Nevertheless, so-called multi-fidelity statistical models can still benefit from TD-DFT as a physics-based regularization to improve generalization to unseen chemistries. TD-DFT calculations still remain orders of magnitude cheaper and more easily automated than experiments, which have a much higher opportunity cost for false positive predictions.This means that the amount of TD-DFT data available, and the ability to acquire more, still far exceeds that from experiments. In this work, we benchmark several approaches for leveraging multi-fidelity combined computational and experimental datasets to predict the optical properties of dye molecules with high transferability to unseen chemical domains. We compare pre-training and transfer learning strategies across several datasets for multiple prediction tasks. Our findings improve the accuracy and generalizability of the machine learning models, enabling researchers to use them more effectively in generative modeling and molecular design.

629340 Multi-Task Property Prediction: Importance of the "Chemist-in-the-Loop" in Model Building

Abstract:

In absence of experimental data, the ability to predict various thermophysical properties related to chemicals of interest becomes crucial in many engineering applications. Among these applications are phase-equilibria calculations, energy balances, and the evaluation of process alternatives [1]. Quantitative Structure-Property Relationship models or "QSPR" models enable such predictions by relating the molecular structure in a machine-readable format (molecular descriptors) to the property of interest through a mathematical model [2].

Recent developments in the field of Deep Learning (DL) and especially Graph Neural Networks (GNN) have eliminated the tedious task of choosing a suitable molecular descriptor for the task at hand, as they can learn an optimal representation from a molecular graph and map them to the target property through backpropagation [3], [4]. Traditionally, models are built to predict one specific property or target. However, DL models can predict several properties simultaneously also known as multitask learning [5]. Here the models might improve their performance through inductive transfer learning i.e. while learning to predict property "A", the model might need less effort to learn how to predict property "B" and might also be able to transfer the newly gained knowledge into the new domain (task) [5]. This is especially relevant in cases where good quality experimental data are scarce [6]. However, an improvement of the model's predictive prowess is not always the case and "negative" transfer might occur [7].

In this work, we will demonstrate that domain knowledge plays an important role in ensuring "positive" learning when dealing with the multi-task prediction of molecular properties by showcasing two case studies: 1) "seemingly" non-related properties in the form of the Gibbs free energy of formation and the acentric factor 2) theoretically related properties in the form of the critical temperature and the acentric factor. For each property, a single task GNN-based model will be developed to serve as a benchmark model to illustrate the potential improvement when applying multi-task transfer learning. The comparative assessment between the two case studies will demonstrate that although AI-based techniques and tools might offer improved results compared to conventional modeling techniques, the "chemist-in-the-loop" [8] is an indispensable element in building and improving AI-based property prediction models.

References

[1] J. Frutiger, I. Bell, J. P. O'Connell, K. Kroenlein, J. Abildskov, and G. Sin, "Uncertainty assessment of equations of state with application to an organic Rankine cycle†," *Mol. Phys.*, vol. 115, no. 9–12, pp. 1225–1244, 2017.

[2] N. D. Austin, N. V. Sahinidis, and D. W. Trahan, "Computer-aided molecular design: An introduction and review of tools, applications, and solution techniques," *Chem. Eng. Res. Des.*, vol. 116, pp. 2–26, 2016.

[3] C. W. Coley, R. Barzilay, W. H. Green, T. S. Jaakkola, and K. F. Jensen, "Convolutional Embedding of Attributed Molecular Graphs for Physical Property Prediction," *J. Chem. Inf. Model.*, vol. 57, no. 8, pp. 1757–1772, Aug. 2017.

[4] K. Yang *et al.*, "Analyzing Learned Molecular Representations for Property Prediction," *J. Chem. Inf. Model.*, vol. 59, no. 8, pp. 3370–3388, 2019.

[5] S. Ruder, "An Overview of Multi-Task Learning in Deep Neural Networks," *arXiv*, no. May, Jun. 2017.

[6] A. M. Schweidtmann, J. G. Rittig, A. König, M. Grohe, A. Mitsos, and M. Dahmen, "Graph Neural Networks for Prediction of Fuel Ignition Quality," *Energy & Fuels*, vol. 34, no. 9, pp. 11395–11407, Sep. 2020.

[7] W. Zhang, L. Deng, L. Zhang, and D. Wu, "Overcoming Negative Transfer: A Survey," *arXiv*, pp. 1–15, Sep. 2020.

[8] T. J. Wills, D. A. Polshakov, M. C. Robinson, and A. A. Lee, "Impact of Chemist-In-The-Loop Molecular Representations on Machine Learning Outcomes," *J. Chem. Inf. Model.*, vol. 60, no. 10, pp. 4449–4456, Oct. 2020.

626631 Specific Ion Effects in Aqueous Electrolyte Solutions from First Principles-Derived Machine-Learning Potentials

Abstract:

We investigate specific ion effects on the structure and dynamics of water in aqueous electrolyte solutions using atomistic machine-learning models constructed from the SCAN approximation of DFT predictions. Ion specificity of bulk electrolyte solutions following the Hofmeister series can induce wide-ranging effects on the dynamics of water - kosmotropic ions mitigate water mobility and chaotropic ions accelerate water molecules. Many existing studies of this phenomena apply conventional empirical models which are inherently limited in transferability and accuracy across a range of concentrations. In this work, in order to explore ion solvation characteristics with a first principles level of accuracy, we construct deep neural network models capable of learning highly complex and multi-dimensional interactions native to DFT representations. We find that these models overcome the limitations of conventional empirical models in representing water dynamics with concentration dependence. We then use the models to probe the underlying mechanisms of ion-induced water structure and mobility for a series of alkali halide ions (KCl, CsCl, NaBr, NaCl) in bulk solution.

624925 Combining Molecular Dynamics Simulations and Active Learning to Study the Hydrophobicity of Chemically Heterogeneous Surfaces

Abstract:

The hydrophobicity of a material is a key property that influences behavior in aqueous environments. While hydrophobicity can be quantified experimentally and predicted computationally for small molecules and larger, chemically uniform surfaces, predicting the hydrophobicity of chemically heterogeneous surfaces – surfaces with polar and nonpolar groups spatially arranged in different patterns on nanometer length scales – remains an outstanding challenge. While classical approaches consider additive effects of various chemical groups on the hydrophobicity based on their area coverage, these relations often fail to capture trends in the hydrophobicity of chemically heterogeneous surfaces with the same mole fractions of different chemical groups but vastly different hydration responses. Probing chemically heterogeneous systems using experimental methods is also challenging due to the length scale of the heterogeneity (on the order of nanometers). Instead, computational methods like traditional molecular dynamics (MD) simulations and enhanced sampling techniques have made progress in understanding the effect of chemical heterogeneity on hydrophobicity. However, two major challenges in using computational approaches to explore the hydrophobicity of chemically heterogeneous surfaces are the large sampling times required for enhanced sampling techniques and the vast number of possible spatial patterns and combinations of polar/nonpolar groups possible in even small molecular systems.

In this work, we present a hierarchical active learning approach to efficiently study the effect of patterning of polar and nonpolar groups on the hydrophobicity of self-assembled monolayers (SAMs). We employ atomistic molecular dynamics simulations and indirect umbrella sampling (INDUS) to label each chemically heterogeneous SAM with a hydration free energy (HFE). We then combine INDUS simulations and 3D Convolutional Neural Networks (CNNs) in an active learning framework. The active learning framework uses a Gaussian Process Regression model to predict the HFE of a SAM and an associated uncertainty in the HFE prediction based on the pattern of polar and nonpolar groups embedded on the SAM. The active learning model combines the labels and their associated uncertainty into an acquisition function which is designed to efficiently explore regions of pattern space with high uncertainty and find patterns which show large deviations from ideal behavior. Using the hierarchical active learning approach, we explore the relationship between patterning and hydrophobicity and recreate an "envelope" of HFEs associated with various mole fractions of the polar component. We also identify patterns with large positive and negative deviations from a linear combination of HFEs at the given mole fraction. Comparison of the HFE envelope for different polar end groups further reveals how deviations from a linear combination of HFEs depend on the end group chemistry. Finally, we analyze structural metrics of interfacial water near the SAM surface to understand how chemical context plays a role in determining hydrophobicity and why some motifs on patterned SAMs are more hydrophobic than others.

622461 Machine Learning the COSMO Model for Predicting Thermodynamics of Electrolyte Mixtures

Abstract:

Bottom-up design of electrolyte mixtures for battery systems requires predicting macro thermodynamic properties from molecular constituents. Molten salt electrolyte batteries require conditions far above room temperature to operate, but have desirable chemical stability, motivating their potential use in aerospace energy storage applications. Therefore, discovering mixtures with increasingly lower eutectic melting points is desirable. A model that can approximate chemical activity is a valuable tool to search through the vast compositional design space. Machine learning can predict properties of materials such as vibrational free energies, electronic energy gaps, and thermal conductivities. Moreover, they can learn physical models such as interatomic potentials.

The COSMO-SAC model uses theory and empirical parameterization to predict liquid-vapor and liquid-solid properties using first-principles calculations. However, obtaining activity coefficients required for parameterizing the COSMO-SAC model is costly and limited to a select chemical space. In this work, we explored how machine learning methods could learn the COSMO-SAC model and bridge density functional theory calculations to liquid phase thermodynamic properties. Our data-driven approaches use existing databases for sigma profiles of organic solvents. Our models attempt to learn using the data of different fidelity to learn activity coefficients of mixtures. First, an optimal machine learning model is constructed for each dataset. Our machine learning algorithms use the sigma-profile as an input feature to predict binary mixtures' activity coefficients using support vector machines. Each dataset uses different choices of functionals, methods, and basis sets. Therefore, our ensemble model attempts to predict corrected activity coefficients given the combination of all the model outputs. The activity coefficients used for training are generated using the COSMO-SAC model. This approach enables the extraction of meaningful information from existing datasets to improve the COSMO-SAC model for obtaining thermodynamic properties of electrolyte mixtures. With the liquid phase activities, we can identify electrolyte mixtures that meet desired phase equilibria conditions.

623156 Generating 3D Conformer Ensembles of Molecular Graphs with End-to-End Differentiable Networks

Abstract:

Prediction of small molecule 3D conformer ensembles plays an important role in many areas of cheminformatics, from the calculation of accurate reaction rates to the drug discovery pipeline. Traditional distance geometry-based generators have several issues including the need for computationally-expensive energy minimization, over-parameterization of the final structure, poor coverage of the true conformer distributions (i.e. lack of diversity in the generated ensembles), and iterative and non-deterministic treatment of tetrahedral chiral centers.

We propose a novel machine learning approach to generate distributions of low-energy 3D conformers directly from molecular graphs. Our approach comprises three steps. First, we predict the local 3D structure of all non-terminal atoms, which we deem local environments, by combining self-attention layers and message passing neural networks. These predictions correspond the the bond distances and bond angles of the final conformer. Next, we predict the dihedral angles that join all pairs of neighboring local environments. Importantly, because we know all bond angles of local environments, we only predict a single torsion angle value, which allows us to recover all the dihedral angles between local environments. That is, we develop a canonical representation of dihedral angles, which allows us to predict exactly the number of degrees of freedom of the conformer. Finally, we assemble all predicted pairs of local environments together to construct the conformer.

Since we build the conformer from predictions of local environments, we can deterministically handle tetrahedral chiral centers without the need for iterative optimization. To promote diverse conformer ensembles, we train our network end-to-end using a loss function inspired by optimal transport. Empirically, we outperform popular baselines, showcasing the benefit of our model. Finally, we discuss possible extensions and applications of our differentiable 3D structure generator to downstream tasks such as modeling drug-target interactions.

628693 Graph Invariants As Rapid Features for Classification and Regression of Bulk Hydrogen Intercalation Energies in Reducible Metal Oxides

Abstract:

Intercalating hydrogen into reducible metal oxides drastically alters the physical and chemical properties of the host oxides. These changes can be directly quantified with quantum chemical calculations like density functional theory (DFT). However, due to the existence of many intercalation sites, this approach becomes tedious, requiring significant computational resources. Machine learning (ML) methods have been applied to these systems to accelerate property predictions, such as intercalation energies, band gaps, and crystal structures. Additionally, ML methods can help interrogate the underlying physics that govern the relationships between intercalation and material properties. The features for these models can be tabulated elemental properties (e.g., electronegativity), making the application of the models general and rapid. Alternatively, DFT-derived quantities, such as electronic band centers and charges, have also been used as features with the additional computational cost of a relatively small number of DFT calculations. Although the use of tabulated features can save computational cost, the derived models often require many features and potentially complex ML architectures for increased accuracy. This may obscure the physical interpretation that could be extracted from simple models that use few descriptors.

In this talk, we will discuss our recent application of graph fingerprints, commonly used in convolutional neural networks (CNNs) for image recognition and molecular ML applications, to predict DFT-calculated hydrogen intercalation energies in bulk metal oxides. The graphs represented the lattice structures of the metal oxides, considering the number of bonds and neighbors at each atom center. Specifically, we considered the graphs as Laplacian matrices of the bulk metal oxides and their intercalated counterparts. Instead of using the full matrices, as is typical for CNNs, we collapsed the full matrices into various single-valued graph invariants. These invariants were then used as features for our ML models that comprise a direct physical link to the geometries of the metal oxides. We demonstrate that a simple logistic regression model can be used to classify metal oxides as either intercalating or non-intercalating oxides. We also show that this classification model can be used to focus on a region of training and improve the performance of regression models that predict the intercalation energy of hydrogen in metal oxides. In summary, this work illustrates a quantitative link between the graph invariants representing metal oxide lattice geometries and changes in material properties associated with bulk hydrogen intercalation. Overall, our ML approach can be applied to rapidly screen energy storage materials and assess the stability and performance of electrocatalysts.

630860 Something Old, Something New: Generative Adversarial Approaches to Conditional Sampling in Complex System Simulations

Abstract:

Sampling from the equilibrium distribution of the phase space of atomic-resolution models of molecular systems (and other complex systems modeled by stochastic differential equations) is a crucial step in many workflows for drug discovery, materials science, and more. These workflows require the computation of free energy differences and ensemble averages with respect to the equilibrium distribution. Such calculations are usually accomplished using molecular dynamics simulations together with enhanced sampling techniques that often require the ability to sample from the equilibrium distribution conditioned on the values of some order parameters of the system (e.g., dihedral angles, principal components, etc.) as an intermediate step. Meanwhile in the field of Machine Learning, generative adversarial networks (GANs) have demonstrated state-of-the-art capabilities in generating samples from probability distributions that are concentrated in lower-dimensional manifolds embedded in high-dimensional ambient spaces [1]. For example, GANs have recently been successful in creating images that look like real photographs of human faces in which the GAN creates photos with unique, detailed, and realistic subtle features that are consistent with the coarse features that make up a human face [2]. In this work, we present a computational technique that combines Physics-based biasing methods for sampling conditional distributions with deep learning-based conditional generative adversarial networks (cGANs) [1, 3]. Extending GANs, cGANs sample from conditional distributions, and cGANs have recently appeared in the context of molecular modeling to interpolate continuous molecular trajectories and configurations between two points in simulation time [4]. On one hand, cGANs have the advantage of providing a generative model for the target distribution that does not suffer from the slow mixing times that afflict traditional biasing simulation methods. On the other hand, cGANs are typically trained offline and require large datasets. Physics-based simulation methods, by contrast, rely solely on a model of the system and the dynamics it undergoes without requiring any other prior knowledge. We put forward an approach that couples cGANs with biased Physics-based simulations; we believe this can bring out the best aspects of both techniques, allowing us to sample more effectively. Additionally, we explore the relationship between cGANs and flat-histogram methods [5, 6, 7, 8].

[1] I. J. Goodfellow, J. Pouget-Abadie, M. Mirza, B. Xu, D. Warde-Farley, S. Ozair, A. Courville, and Y. Bengio. Generative Adversarial Networks. arXiv:1406.2661 [cs, stat], June 2014. arXiv: 1406.2661.
[2] Tero Karras, Samuli Laine, and Timo Aila. A style-based generator architecture for generative adversarial networks, 2019.
[3] M. Mirza and S. Osindero. Conditional Generative Adversarial Nets. arXiv:1411.1784 [cs, stat], November 2014. arXiv: 1411.1784.
[4] Hythem Sidky, Wei Chen, and Andrew L. Ferguson. Molecular latent space simulators. Chem. Sci., 11:9459–9467, 2020.
[5] S. Kumar, J. M. Rosenberg, D. Bouzida, R. H. Swendsen, and P. A. Kollman. The weighted histogram analysis method for free-energy calculations on biomolecules. I. The method. Journal of Computational Chemistry, 13(8):1011–1021, October 1992.
[6] B. Roux. The calculation of the potential of mean force using computer simulations. Computer Physics Communications, 91(1-3):275–282, 1995.
[7] P. E. Jacob and R. J. Ryder. The Wang-Landau algorithm reaches the flat histogram criterion in finite time. The Annals of Applied Probability, 24(1):34–53, February 2014.
[8] Z. Tan, J. Xia, B. W. Zhang, and R. M. Levy. Locally weighted histogram analysis and stochastic solution for large-scale multi-state free energy estimation. Journal of Chemical Physics, 2016.

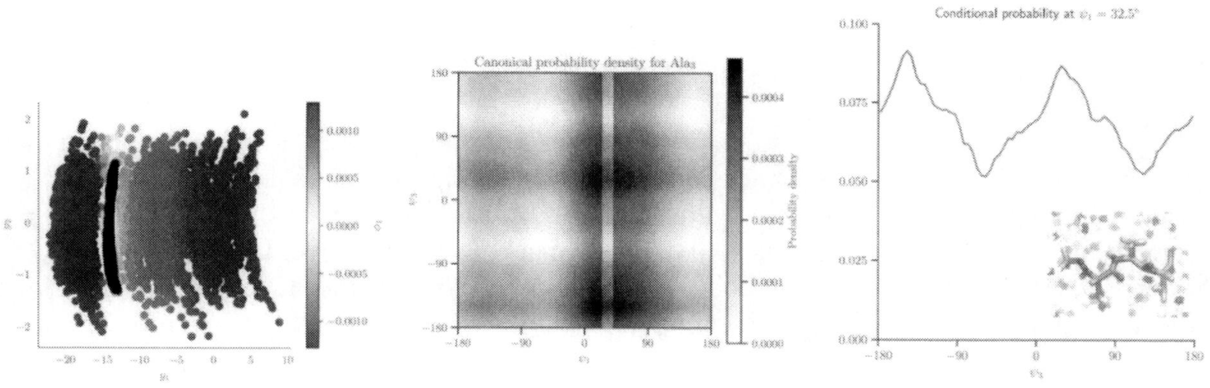

(a) (b) (c)

FIGURE 1. **(a)** conditional sampling (black) of an SDE biased by the order parameter represented by ϕ_1 (color), **(b)** Equilibrium distribution of trialanine (gray) with a highlighted conditional distribution at a specific value of one of its order parameters (orange). **(c)** Depiction of the solvated trialanine molecule (inset) over the conditional probability at the specific value of the order parameter.

621410 Teaching Process Data Analytics and Machine Learning

Abstract:

Teaching chemical engineers to be effective in data analytics is as important today than in any time in the history of the discipline, and advances in sensor technologies, machine learning, and associated software have enabled applications that were not formerly possible. This abstract describes experiences with teaching process data analytics and machine learning over the last few years, including in (1) a new chemical engineering elective course in which the students come from chemical engineering, mechanical engineering, and engineering management, (2) a professional education course cotaught by chemical and mechanical engineering faculty whose learners primarily come with chemistry and chemical engineering backgrounds and many years of industrial experience, and (3) an undergraduate chemical engineering concentration in process data analytics with courses coming from a mixture of multiple disciplines including chemical engineering, computer science, economics, management. The overview will describe the challenges in teaching data science to chemical engineers and scientists and engineers working in the chemical and biotechnology industries, and strategies for overcoming those challenges.

622120 Teaching Big Data Science to Undergraduate Students in the University of Iowa

Abstract:

In this presentation, we will provide an overview of our course development and our experiences in teaching big data science and analytical skills to the undergraduate students in chemical engineering in the University of Iowa. These courses include: (a) Machine Learning (ML) and Artificial Intelligence (AI) course; (b) statistical analysis course; and (c) thermodynamic course. They are designed to be offered to students from freshman to junior years. We found that in general, the chemical engineering students are not getting used to use programming to analyze the data, in part because many courses they are taking focus more on the lab experiment while the analysis of the lab results is elementary and often done in Excel or by calculator. However, the students are willing to learn new analytical skills, and provided with well-designed teaching materials and problems, they can grasp the big data skills quickly. While the outcome of our experience has been encouraging, we also found there is clear lag in the field of chemical engineering to train our next generation with big data tools. This lag is part is due to the institutional tradition that emphasizes on the lab experiments and often deterministic way of thinking (with less consideration of probabilities), and in part due to the slowness in re-design the curriculum to reflect new development and ease access of ML and AI. We will provide examples to support the findings above, share the lessons we learned, and provide thoughts for next steps.

623878 Textile: Tutorials in Experimentalist Interactive Learning

Abstract:

Background:

Confocal microscopy is a common preclinical imaging modality to study a range of tissue features. Fluorescent image microscopy can provide insights into cell density, morphology, and cell-cell relationships. In our lab, we use immunofluorescent staining to investigate changes in cell and microenvironment features in response to development, injury, and treatment in the brain. We have developed image processing pipelines that threshold, segment, and quantify cell features. More recently, we have incorporated data science-enabled analysis and techniques into our imaging workflow. While the lab is comprised of researchers who are wet-lab experimentalists and data-driven analysts, only the second group has formal data science training. Wet-lab experimentalists can significantly benefit from having data science literacy to apply data science methodologies to their imaging datasets. However, there is no formal training regime to train wet-lab experimentalists in lab-specific data science methodologies. To address this training gap, we developed Tutorials in EXperimentalisT Interactive LEarning (TEXTILE) for data-science enabled fluorescent image processing.

TEXTILE is a module-based, semi-linear educational system developed with laboratory-based learning in mind. Modules combine data science fundamentals with specific biological imaging processing techniques. For example, we pair a data science module about data management with an image processing module on quantitative cell morphology characterization. The modular format allows us to train students on managing experimental data and pair that with quantifying features of cells that are relevant to their specific project of interest. Additionally, we designed TEXTILE such that the learning process is adaptable to starting skill level. Learning paths are based on entry knowledge but allow students to pick and choose their modules based on interest or previous experience. For example, someone with ample coding knowledge can skip the introductory Python modules and begin with the experimental image processing modules. Finally, we designed TEXTILE for a variety of degree levels. As a laboratory at a large public university, we engage with undergraduate and high school students, in addition to master's thesis and doctoral thesis students. We accomplished degree-level accessibility through the semi-linear system that assumes no prior knowledge in our research area. Since we already use TEXTILE to train high school students participating in laboratory research that have no previous knowledge of the subject topic, we aim to include more local high school students and expand TEXTILE for science education and outreach as well as in-laboratory education.

Methods:

TEXTILE is composed of topic-specific modules. Each module is designed around a data science topic, lasts an hour and a half, and contains four sections: the pre-module activity, main module, post-module reflection, and feedback form (Figure 1a). The instructor sends the pre-module activity to students for self-driven learning a week before teaching the main module. The pre-module's purpose is to prime students for the module's main lesson with a high-level activity. For example, in our data management module, students analyze a grocery shopping trip and decide where to store specific categories of food and perform a space evaluation on how much room in each storage location the food will take up.

The main module aims to teach the bulk material in digestible lectures broken up by interactive sessions. The main module's detailed structure consists of six parts: introduction presentation, interactive session one, main presentation workshop, interactive session two, concluding presentation, and a reflection session (Figure 1b). The introduction presentation introduces the module's topic, outlines the structure the main module will follow, and leads into the first interactive session. Interactive session one guides the student through an exploration of the module material. For coding-based modules, instructors lead interactive sessions with Jupyter Notebook. During interactive session one, the instructor walks students through the code's basics as students code along. After the students have written the code, the instructor begins the main presentation, which has two purposes: (1) to increase the students' connection between the pre-module activity and the module introduction and (2) to dive deeper into the technical aspects of the first interactive session. Once students understand the technical elements of the main module, they begin interactive session two, where they apply technical details of the lesson to further explore the code written in the earlier interactive session. After students have an opportunity to explore, the instructor gives the concluding presentation, which includes wrapping up the main takeaways, emphasizing the module's motivation, and thanking students for their participation.

Finally, the main module ends with a prompt for students to reflect on what they have learned and how their opinions about the module's topic have changed since the pre-module activity. The post-module reflection is completed individually by each student a day or two after the main module. The purpose is to provide a student-led review activity that encourages them to explore content from the main module on their own time to solidify concepts and build confidence. After the whole module is completed, students are sent a feedback form through Google Forms. With the feedback form, students reflect on their growth and the impact of each component of the module. The feedback is then implemented back into the module for the next cohort of students.

Results:

In the summer of 2020, the first cohort of students went through six TEXTILE modules of paired data science and image processing techniques used within the lab. The first cohort had 15 students: six high school students, four undergraduates, and five graduate students. The first cohort was taught live by a Ph.D. student from the lab's data-driven analysis team. The topics covered included GitHub and version control, introduction to data management, introduction to experimental data management specific to the lab, intro to image processing, thresholding, image segmentation, cellular morphology, and quantifying cell shape features. By the end of the first cohort, all students completed every TEXTILE module and the high school and undergraduate students developed individual projects using protocols learned over the summer. They each created Jupyter Notebooks for a self-led project with a brain cell morphology lit review, data management plan, and image processing pseudo-code. Many students from the summer 2020 cohort chose to continue researching with the lab and are currently developing these projects further.

In fall 2020, a small cohort of two undergraduate students explored the TEXTILE modules asynchronously. The students watched a recording of the main module and participated in asynchronous discussions with previous students and a Ph.D. student-instructor via Slack. We also provided these students with published image data that no one previously analyzed for cell morphology features. The students are currently completing retroactive analysis on this data using techniques learned from various TEXTILE modules.

For summer 2021, TEXTILE is expanded to include three modules on machine learning and two additional modules on experimental image acquisition and manual cell counting, incorporating student feedback from 2020. The main takeaways from student feedback included developing a formalized mentoring process for student projects and teaching via a hybrid live and asynchronous format.

Conclusions:

TEXTILE is a successful data science educational system for teaching data science techniques commonly used in experimental research labs. The TEXTILE system works for protocol training of fellow lab members, onboarding new students, and science education outreach. The combination of interactive sessions with instructor-led presentations gives students ample opportunity to work hands-on with real data from the lab in a guided environment. Students from the first cohort had the opportunity to develop their cell morphology research projects. In contrast, students from the second cohort took the lessons learned from TEXTILE and applied them for retroactive analysis of previously published data. The TEXTILE system is built to be robust and adaptable so that other laboratories can follow the methodology to develop training modules for their own data science techniques.

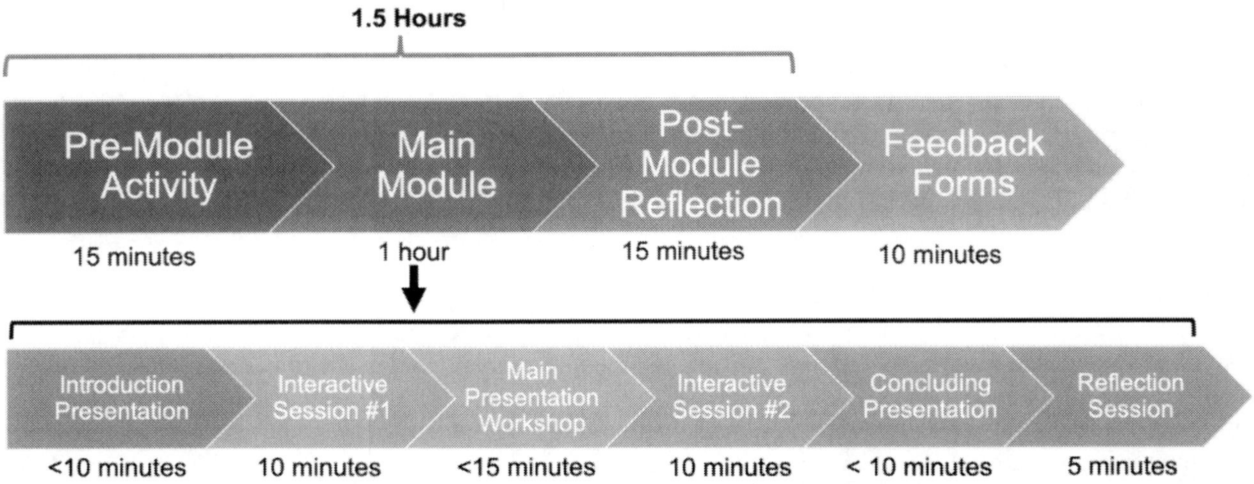

Figure 1. (a) Top row: TEXTILE process composed of four parts – pre-module activity, main modules, post-module reflection, and feedback forms. (b) Bottom Row: The six sections of the main module - introduction presentation, interactive session #1, main presentation workshop, interactive session #2, concluding presentation, reflection session.

622481 Teaching Artificial Intelligence to Chemical Engineers: Experiences from a 35-Year-Old Course

Abstract:

Given the recent success of artificial intelligence (AI) in a wide variety of applications, it has become quite clear that chemical engineering students need to become familiar with applied AI techniques [1]. There is great interest in the academic community in developing a course to address this important need. I have been teaching such a course to chemical engineering undergraduate and graduate students since 1986 [2], first at Columbia for three years, then at Purdue for about twenty years, and back again at Columbia now. It was mostly on *symbolic AI* in the early years, with an emphasis on *knowledge-based expert system* concepts and techniques, for various applications in chemical engineering such as design, diagnosis, control, optimization, and process safety. From the 1990s onwards, as machine learning techniques started showing promise, my course evolved to include topics such as clustering, neural networks, statistical classifiers, graph-based models, and genetic algorithms. The current version at Columbia has the first one-third of the course on *symbolic AI* and the rest on machine learning-based *numeric AI* concepts and techniques. It is mainly a project-based course where the students develop systems using AI techniques for chemical engineering applications. In the early years, the students programmed in AI languages such as LISP and OPS5, but now using the easier tools such as Python and TensorFlow. In this presentation, I will discuss my experiences in teaching this course, including the syllabus and the projects we have done over the decades.

Reference:

1. Venkatasubramanian, "The Promise of Artificial Intelligence in Chemical Engineering: Is it here, finally?", *Perspective article, AIChE Journal,* 65(2), pp. 466-478, 2019.

- Venkatasubramanian, V. "Artificial Intelligence in Process Engineering: Experiences from a Graduate Course," *Chemical Engineering Education*, Fall Graduate Education Issue, November 1986, pp. 188 – 192.

621552 Deriving Descriptors and Physical Understanding from Data in Catalysis Research

Abstract:

Important properties of a catalytic material, such as activity and selectivity, are in general difficult to predict, in particular from first principles. The problem lies in the extreme complexity of the relation between the atomic composition of a catalyst and its catalytic properties at realistic temperatures and pressures. We demonstrate how to bridge this complexity with artificial intelligence (AI).

Single-atom metal alloy catalysts (SAACs) have recently become a very active new frontier in catalysis research. The simultaneous optimization of both facile dissociation of reactants and a balanced strength of intermediates' binding make them highly efficient and selective for many industrially important reactions. However, discovery of new SAACs is hindered by the lack of fast yet reliable prediction of the catalytic properties of the sheer number of candidate materials. In this work, we address this problem by applying a compressed-sensing data-analytics approach SISSO [1] parameterized with density-functional inputs. Our approach is faster and more accurate than the current state-of-the-art linear relationships. Besides consistently predicting high efficiency of the experimentally studied Pd/Cu, Pt/Cu, Pd/Ag, Pt/Au, Pd/Au, Pt/Ni, Au/Ru, and Ni/Zn SAACs (the first metal is the dispersed component), we identify more than two hundred yet unreported candidates [2]. Some of these new candidates are predicted to exhibit even higher stability and efficiency than the reported ones. Our study demonstrates the importance of breaking linear relationships to avoid bias in catalysis design, as well as provides a recipe for selecting best candidate materials from hundreds of thousands of transition-metal SAACs for various applications.

Moreover, using subgroup discovery, an AI approach that discovers statistically exceptional subgroups in a dataset, we develop a strategy for identification of most important parameters of a catalytic material and competing mechanisms of a catalytic reaction. The approach is used to develop physical understanding of hydrogen activation at SAAC's [2] and address the problem of converting carbon dioxide (CO_2) to fuels and other useful chemicals [3]. Large-scale conversion would alleviate detrimental effects of CO_2 excessive emissions on the environment. The AI model is trained on high-throughput first-principles based data for a broad family of oxides. We demonstrate that an electron transfer to the pi*-antibonding orbital of the adsorbed molecule and the associated bending of the gas-phase linear molecule, previously proposed as the indicator of activation, are insufficient to account for the good catalytic performance of experimentally characterized oxide surfaces. Instead, our AI approach identifies the common feature of these surfaces in the binding of a molecular O atom to a surface cation, which results in a strong elongation and therefore weakening of one molecular C-O bond. This finding suggests to use the C-O bond elongation as an indicator of CO_2 activation. Based on these findings, we propose a set of new promising oxide-based catalysts for CO_2 conversion, and a recipe to find more.

[1] R. Ouyang, S. Curtarolo, E. Ahmetcik, M. Scheffler, and L. M. Ghiringhelli, Phys. Rev. Mater. 2, 083802 (2018).
[2] Z.-K. Han, D. Sarker, R. Ouyang, A. Mazheika, Y. Gao, and S. V. Levchenko, Nature Comm. 12, 1833 (2021).
[3] A. Mazheika, Y. Wang, R. Valero, L. M. Ghiringhelli, F. Vines, F. Illas, S. V. Levchenko, and M. Scheffler, arXiv:1912.06515 [cond-mat.mtrl-sci].

Inverse Design of Nanoparticles Using Charge Transfer Properties and Multi-Target Machine Learning

Sichao Li[1] and Amanda S. Barnard*[1]

[1]*School of Computing, Australian National University, Acton 2601 ACT, Australia*

Abstract

We present a new approach to inverse design that uses sets of properties to predict a unique nanoparticle structure, and is based on established multi-target regression and reliable forward structure/property prediction. Feature selection is used to focus the model on the most important characteristics before inverting the problem and simultaneously predicting multiple structural features of a single nanoparticle. The workflow is general, as demonstrated on two nanoparticle data sets, and can rapidly predict property/structure relationships to guide further research and development without the need for additional optimisation or high-throughput sampling.

1 Introduction

Inverse design[1, 2, 3] that prescribes a structure is a primary objective in materials informatics, and the the ultimate goal of much academic and industrial research, but the majority of materials informatics uses machine learning (ML) to make forward predictions of a property of a material (the target label) based on the structural characteristics (the features). Unlike structure/property relationships used to inform synthesis and processing strategies, inverse design involves inverted property/structure relationships, predicting the structural features based on the target property labels. Inverse design is particularly complicated in nanomaterials design, where the finite sizes and multitude of shapes mean the design space is larger.[4]

Conventional approaches to inverse design involve a combination of forward structure/property predictions, optimization and screening, based on high throughput data generation. For example, Tominaga *et al.* proposed an efficient procedure based on genetic algorithms (GAs) for optimisation of a single inverse system [5]. Yang *et al.* also developed a package for inverse design, named IM^{2}ODE (Inverse design of Materials by Multi-Objective Differential Evolution), based on multi-objective differential evolution. These approaches have some challenges. Although they implement structural optimization, and duplicate elimination to increase the efficiency, it is still an optimization method striving to solve a global searching task which is time-consuming. As a result of using real-valued number instead of genes, differential evolution tends to be more effective than GAs, but maintaining a viable population of candidate solutions and low convergence properties are problematic [6, 7]. Alternative approaches using evolutionary algorithms have been reported, but similarly suffer from poor scaling with respect to complexity and often require very expensive fitness functions for evaluation, often taking days to complete. The stopping criteria of genetic algorithms for predicting multi-targets is

*Corresponding author: amanda.s.barnard@anu.edu.au

also unclear [8], and the reliability of the throughput systems generating the data depends on the number and types of competing candidate structures, and can result in impossible compounds [1, 9, 10]. Finally, traditional screening-based approaches can only choose from candidates that are in the data set[11, 12, 13], meaning this approach also suffers from too much specificity. In some cases a single structural characteristic might be enough to guide the selection of a (nano)material, such as the strongly size-dependent optical emission of quantum dots, but in many cases the underlying structure/property relationships are more complicated and so the property/structure relationship will not be useful unless we can extract a more holistic structural profile.

An alternative approach to inverse design is possible in cases where more than one property label is known is to use multi-target regression to predict one profile for a given set of conditions. This eliminates the need for optimization to find a combination of structural characteristics that will work. If enough labels are available, multi-target forward predictions can be inverted, using properties as inputs to predict a target structural characteristics. This overcomes the issue of having insufficient known variables to predict an unknown structure, and the fact that one structural characteristic may be not be sufficient to guide experiments. Here we present a new inverse design workflow using multi-target random forest regressors, and test it on a set of silver nanoparticles. As we will show, by informing the inverse models with information on the important structural characteristics in a conventional (forward) structure/property relationship, new property/structure relationships can be predicted with comparable performance in terms of accuracy and generalisability.

Methods

The data set contains 425 silver nanoparticles[14] originally simulated using electronic structure methods and characterised numerically using statistical analyses to define structural features and electron transfer properties [15, 16, 17]. The repository for this set contains a table of extracted features, which there first cleaned to improve model performance. This set contains face centred cubic and twinned structures with diameters ranging from 0.5 nm to 4.9 nm, and various morphologies defined by zonohedrons enclosed by {100}, {110}, {111}, {210}, {113}, {331}, and {123}. The entire data set can be accessed online [14], along with a comprehensive table of features. Following download outliers were removed, constant features were ignored, and irrelevant (or minimally relevant) features were filtering based on the standard deviation of each feature, as they contribute nothing to the model accuracy. Instances where `IP`<5 eV, `EA`>2.5 eV, `EG`<2.5 eV, -4.05<`EF`<-3.5 eV and `Formation_E`<0.8 eV, were omitted resulted in a final set of 414 nanoparticles. Similarly, strongly correlated features are redundant as they both contribute the same information during learning, and unnecessarily complicate the model by introducing bias that should be reduced. In our study, linearly correlated features are identified using a correlation matrix and features with over 95% correlation are removed and features with a standard deviation of less than 0.1% were eliminated to reduce noise. The data was then scaled using the MinMax scaler and normalized. The remaining set was recursively optimised until the most informative features are retained as described below. The labels included in the data set include electron charge transfer properties and a measure of thermodynamic stability. These are the formation energy (`Formation_E`), the ionization potential (`IP`), the electron affinity (`EA`), the electronic band gap (`EG`) and the energy the Fermi level (`EF`), all measured in eV. Labelling was not conducted as part of this study, and information on how the labeling was conducted can be obtained from references listed on the data repository.

The selection of our machine learning algorithm has been informed by the high-dimensional numerical target labels, which require a multi-target regression model. Borchani *et al.*[18, 3]

categorised existing methods for multi-output regression into problem transformation methods and algorithm adaptation methods. Random forest (FR) is an ensemble technique based on decision trees and the bagging (or bootstrap aggregating) technique that can be used to fit high-dimensional target labels simultaneously with high accuracy. A decision tree predicts the value of a label following decision rules inferred from the features and bagging randomly samples the training set with replacement, reducing the variance. Constructing a large number of decision trees with random feature selection methods such that each node splits on a random selection of the features grows into a random forest, in which the decision of the individual tree is counted to output the mode of the classes for classification or average prediction for regression [19, 20, 21]. Features are ranked during training based on variance reduction during regression tasks, increasing the diversity and avoiding over-fitting. This machine learning method can intrinsically handle multi-task problems since the leaf nodes can refer to any collection of relevant labels. To extend the traditional single-target ensemble predictor to solve multi-target RF learning problems, users can simply substitute the typical univariate trees for multivariate trees, where leaf nodes refer to multiple classes or target labels [22, 23, 24, 25].

The entire workflow involves both forward and inverse model retaining, to improve performance and provide some septicity. A forward RF model is optimised and trained using k-fold cross validation, which enables the regressor to rank the features by Gini importance. Recursive feature elimination was used to select the most important features sufficient for the model to simultaneously predict the property labels without lost of accuracy or generalizability. This is important to reduce the number of predictable structural characteristics to be commensurate with (or similar to) the number of available properties, and while not essential, it is good practice to repeat the forward model optimization and training using only the final subset of features to quantify any loss. The data set is then inverted; the reduced set of important features becoming structural "meta-labels", and the multiple property labels becoming the new "meta-features". The inverted set is then split into training, testing and validation sets, and the optimization, training and validating processes of the multi-target RF model is repeated. The forward model and the inverse model use the same inputs and targets, but in reverse.

Results

Following cleaning the data sets were split into 80% training set and 20% testing set prior to optimization. The same train/test split, with the same random seed, was used for all models. For each regressor we defined a grid of hyper-parameter ranges and searched 1000 random combination of values sampled from the grid, performing 5-fold cross validation at each iteration. All optimized regressors were trained with 10-fold cross validation. The accuracy and generalizability of all of the hyper-parameters-tuned models was assessed using the mean square error (MSE) and the mean absolute error (MAE) to evaluate the training, testing and cross-validation scores. [26] These results are tabulated, or included in learning curves which also quantify any under-fitting or over-fitting.

Forward Prediction

To begin, single-target forward models where trained with 48 retained features following cleaning to predict each target property label individually. These results are captured in Table 1, and provide a benchmark for comparison of the multi-target forward and inverse models. The forward multi-target prediction training was initially conducted with 48 retained features following cleaning and five property labels: `Formation_E`, `IP`, `EA`, `EG`, and `EF`. The results are provided in Table 1, and the learning curve in Figure 1(a). The feature importance

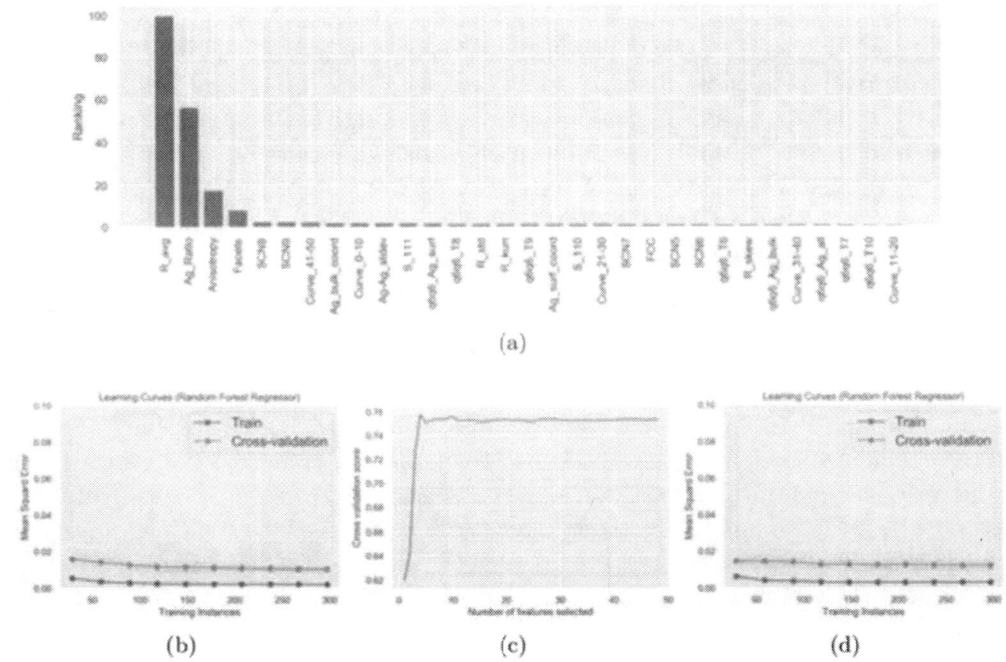

Figure 1: Results for the forward multi-target regression simultaneously predicting (a) the feature importance profile rankings, (b) the learning curve, (c) results of recursive feature elimination showing optimal results can be achieved with 4 features, and (d) the learning curve for forward prediction using only the 4 important features, showing no loss of accuracy and almost no loss of generalisablity. The testing results are evaluated using the mean square error (MSE).

ranking is shown in Figure 1(b), and the recursive feature elimination profile in Figure 1(c) reveal that an optimal model can be achieved using only 4 structural features: **Ag_Ratio**, **R_avg**, **Anisotropy** and **Facets**. The results using this reduced feature set is provided in Table 1 showing minimal loss of performance, and the associated learning curve is shown in Figure 1(d). Retraining the forward multi-target model with only important features once again reduced the performance by less than 1%, and the results of both models compare well with the associated single-target predictions, as shown in Table 1. The performance of the single-target prediction for the energy of the Fermi level (**EF**) is an obvious outlier, with larger errors than the others, which is consistent with previous results using this data set,[16] due to the bimodal distribution of this label.

Table 1: Forward predictions using the multi-target regression model of silver nanoparticles, with all of the structural features retained after data cleaning, or the top 4 important features identified using recursive feature elimination. The testing results are evaluated using the mean absolute error (MAE) and the mean square error (MSE). Noting the data has been standardized between 0 and 1, these MAE and MSE are directly comparable.

Prediction	Feature Set (Number)	Target (Number)	MAE	MSE
Forward	Retained (48)	Formation_E (1)	0.0247	0.0011
Forward	Retained (48)	IP (1)	0.0952	0.0151
Forward	Retained (48)	EA (1)	0.0579	0.0061
Forward	Retained (48)	EG (1)	0.0338	0.0027
Forward	Retained (48)	EF (1)	0.1486	0.0386
Forward	Retained (48)	All (5)	0.0614	0.0090
Forward	Important (4)	All (5)	0.0710	0.0120

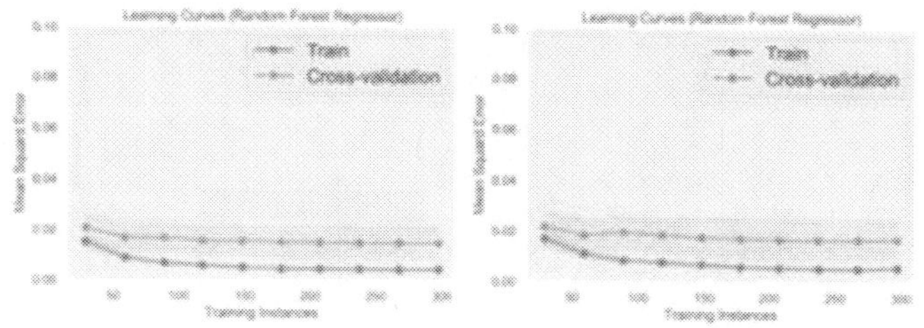

(a) (b)

Figure 2: Learning curves for (a) 5 property meta-features predicting 4 structural meta-labels, and (b) 4 property meta-features predicting 4 structural meta-labels, removing `Formation_E` that is difficult to obtain experimentally.

Inverse Prediction

To train the inverse model the 5 properties become meta-features and the 4 structural features become meta-labels. The multi-target RF model is re-optimised and trained using the procedure above, and the results are provided in Table 2 and Figure 2. In this case the number of facets (an indicator of the nanoparticle morphology) is a difficult label to predict, suggesting an alternative shape-dependent feature/meta-label may improve the outcome. All of the other single-target meta-labels perform similarly to the multi-target prediction. We also reduced the number of meta-features, dropping the formation energy (which is difficult to measure experimentally) and re-optimised and trained the multi-target model. The results in Table 2 show a small loss of performance with the MAE and MSE increasing by only $\sim 0.5\%$, and the learning curves in Figure 2(b) being virtually indistinguishable.

Table 2: Inverse predictions using the multi-target regression model for the silver nanoparticles. The structural meta-labels are listed in order of importance in the forward model. The testing results are evaluated using the mean absolute error (MAE) and the mean square error (MSE). Noting the data has been standardized between 0 and 1, these MAE and MSE are directly comparable.

Prediction	Feature Set (Number)	Target (Number)	MAE	MSE
Inverse	All (5)	All (4)	0.0732	0.0211
Inverse	Electronic (4)	All (4)	0.0788	0.0226
Inverse	All (5)	`Ag_Ratio` (1)	0.0207	0.0032
Inverse	All (5)	`R_avg` (1)	0.0174	0.0008
Inverse	All (5)	`Anisotropy` (1)	0.0910	0.0212
Inverse	All (5)	`Facets` (1)	0.1576	0.0529

To validate the results, three instances have been randomly selected from the testing set, and the predicted values of the structural characteristics are compared with teh ground truth in Figure 3. Here we can see that the inverse model performs exceptionally well, and can be used with confidence. To give an insight into how to use a trained inverse model in practice, we can consider a hypothetical instances with reasonable properties, and predict the nanoparticle that would exhibit them using each of our model. For a silver nanoparticle with [`Formation_E` =0.4, `IP` = 4.0 eV, `EA` = 3.0 eV, `EG` = 2.0 eV, `EF` = 2.5 eV] the inverse model prescribes [`Ag_Ratio`=0.56±0.0443, `R_avg`=22±1.3084 (Å), `Anisotropy`=1.42±0.0848, `Facets`=18±2]. The uncertainties were calculated using the MSE and are useful, as they provide researchers with a fault tolerance; essential for translation into industry.

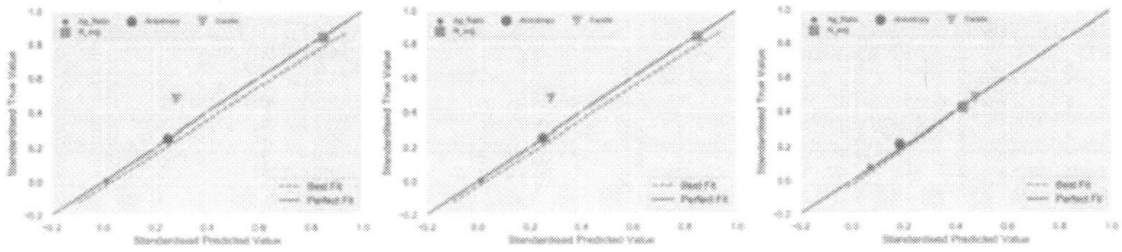

(a) (b) (c)

Figure 3: Examples of the inverse prediction compared to the ground truth for three randomly chosen examples from the testing set.

Discussion

The inverse models predicted here show remarkably good performance (with low MSE and MAE), comparable to forward models using the same reduced sets of important features, and single-target predictions. Although different accuracy metrics are available, since we are using a complicated and highly non-linear regressor (multi-target random forest) the coefficient of determination (R^2) is inappropriate for interpreting dependent variability in our study [27]. Both MAE and MSE are practical in expressing the average model error, and in our study all of the data as standardised and normalised, so there is a consistent range and unit, but this is not enough for evaluation. In addition to this, and for their intrinsic utility, the performance has been evaluated using learning curves. While a disadvantage of this multi-target regression is that the 45 degree plots typically used to compare the predicted and true values cannot be used for more than one target, learning curves for models offer the advantage of reporting both the training accuracy (including under-fitting) and generalisability (including over-fitting) via the cross-validation score, which is important since biased trees can be created if some classes dominate [28, 29]. Although the learning curved presented here indicate we have convergence with respect to the size of the training sets, it is likely that the results could be improved if more nanoparticle properties were available. To achieve a good result we recommend gathering as many properties as possible, and restricting the prediction to only the important structural characteristics, which should be identified using an entirely data-driven approach as we have done here (to help avoid evaluation bias). This step achieves a better balance in the number of known and unknown variables than an unrestricted multi-target model, reduces the computational complexity and the computation time. While it necessitates training an optimised forward model before an inverse model is attempted, it also comes with the added advantage of providing a preview into the overall performance that can be expected from the inverse model, though the forward model is likely to be superior. If a very large number of properties are available it might also be worthwhile cleaning the labels by checking for correlations and noise as we have done for the original feature set, since they will eventually become meta-features. For example, in the case of the inverted predictions in the silver nanoparticle data set the correlation between `Formation_E` and `EG` is 97.4%, which would typically require one to be eliminated if they were among the original features. When we did remove this label/meta-feature as there was small impact on the model performance, but removing too many properties (or retaining too many highly correlated properties) could have more significant consequences. In the present study we also preserved the original dimensions of the properties labels, since random forests are forgiving of data cleanliness and can perform well without standardization or normalization, but this might also be worth investigating if the number of properties is large.

Conclusions

In conclusion, we have demonstrated a new approach to inverse design using multi-target regression based on random forests using a publicly available data set for silver nanoparticles. We have shown that an inverse design model that predicts a set of structural characteristics from a set of properties can have similar performance to a traditional forward model (predicting properties from the structural features) and single-target models. The inverse model benefits computationally from restricting the structural characteristics to only those deemed important in a traditional forward structure/property model using feature importance rankings, and undertaking a precursory forward model also provides a useful indication of the approximate performance that can be expected for the inverse model, though the former is likely to be superior. Overall the inverse design workflow used in the study is general and flexible enough to be applied to other nanoparticles and identify other types of property/structure relationships.

Acknowledgements

Computational resources for this project have been supplied by the National Computing Infrastructure (NCI) national facility under partner Grant p00.

Conflicts of Interest

There are no conflicting interests to declare.

References

[1] Zunger A 2018 *Nature Reviews Chemistry* **2** 1–16

[2] Sanchez-Lengeling B and Aspuru-Guzik A 2018 *Science* **361** 360–365

[3] Schmidt J, Marques M R, Botti S and Marques M A 2019 *npj Computational Materials* **5** 1–36

[4] Barnard A, Motevalli B, Parker A, Fischer J, Feigl C and Opletal G 2019 *Nanoscale* **11** 19190–19201

[5] Tominaga D, Koga N and Okamoto M 2000 Efficient numerical optimization algorithm based on genetic algorithm for inverse problem *Proceedings of the 2nd Annual Conference on Genetic and Evolutionary Computation* pp 251–258

[6] Zhang Y Y, Gao W, Chen S, Xiang H and Gong X G 2015 *Computational Materials Science* **98** 51–55

[7] Rocca P, Oliveri G and Massa A 2011 *IEEE Antennas and Propagation Magazine* **53** 38–49

[8] Reed P, Minsker B and Goldberg D E 2000 *Water Resources Research* **36** 3757–3761

[9] Schleder G R, Padilha A C, Acosta C M, Costa M and Fazzio A 2019 *Journal of Physics: Materials* **2** 032001

[10] Zunger A 2019 Beware of plausible predictions of fantasy materials

[11] Hanakata P Z, Cubuk E D, Campbell D K and Park H S 2018 *Phys. Rev. Lett.* **121**(25) 255304

[12] Wan J, Jiang J W and Park H S 2020 *Carbon* **157** 262–269 ISSN 0008-6223

[13] Ma C, Zhang Z, Luce B, Pusateri S, Xie B, Rafiei M and Hu N 2020 *npj Computational Materials* **6** 40

[14] Barnard A S and Sun B 2017 Silver nanoparticle data set. csiro. data collection.

[15] Barnard A S and Opletal G 2019 *Nanoscale* **11**(48) 23165–23172 URL `http://dx.doi.org/10.1039/C9NR03940F`

[16] Sun B, Fernandez M and Barnard A S 2017 *Journal of Chemical Information and Modeling* **57** 2413–2423

[17] Parker A J and Barnard A S 2020 *Nanoscale Horizons* **5** 1394–1399

[18] Borchani H, Varando G, Bielza C and Larranaga P 2015 *Wiley Interdisciplinary Reviews: Data Mining and Knowledge Discovery* **5** 216–233

[19] Breiman L 2001 *Machine learning* **45** 5–32

[20] Breiman L 1996 *Machine learning* **24** 123–140

[21] Ho T K 1998 *IEEE transactions on pattern analysis and machine intelligence* **20** 832–844

[22] Segal M R 1992 *Journal of the American Statistical Association* **87** 407–418

[23] Segal M and Xiao Y 2011 *Wiley Interdisciplinary Reviews: Data Mining and Knowledge Discovery* **1** 80–87

[24] Kocev D, Vens C, Struyf J and Džeroski S 2007 Ensembles of multi-objective decision trees *European conference on machine learning* (Springer) pp 624–631

[25] Kamiński B, Jakubczyk M and Szufel P 2018 *Central European journal of operations research* **26** 135–159

[26] Razi M A and Athappilly K 2005 *Expert Systems with Applications* **29** 65–74

[27] Spiess A N and Neumeyer N 2010 *BMC pharmacology* **10** 6

[28] Mac Namee B, Cunningham P, Byrne S and Corrigan O I 2002 *Artificial intelligence in medicine* **24** 51–70

[29] Segal M R 2004 Machine learning benchmarks and random forest regression

624822 Design of Optoelectronic Self-Assembling π-Conjugated Peptides Via Combined Experimental-Computational Multi-Fidelity Active Learning

Abstract:

In this work we integrate experimental and computational data streams using multi-fidelity Bayesian optimization and deep representational learning within an active learning platform to accelerate the design and discovery of π-conjugated peptides capable of self-assembling into nanowire-like structures with emergent optoelectronic properties. π-conjugated peptides are tri-block molecules composed of a central π-core flanked by oligopeptide wings demonstrated to self-assemble into pseudo-1D nanoaggregates resembling biocompatible nanowires with potential applications in engineered linkages with living cells for repairing neuronal damage and interfacing with photosynthetic machinery for energy harvesting and storage. The amino acid sequence comprising the oligopeptide wing influences the self-assembly and resultant nanostructure geometry, however exhaustive traversal of possible oligopeptide wings is prohibitively expensive via either simulation or experiment alone due to the combinatorial explosion of the molecular design space. Using graph neural networks and deep representational learning we create a continuous, low-dimensional representational of our discrete, high-dimensional molecular design space that we expose to a multi-fidelity Bayesian optimization platform fusing inexpensive, high-volume, and approximate simulation data with expensive, low-volume, and accurate experimental data to rationally navigate and identify molecules with engineered optoelectronic properties. This active learning search identifies new molecules that are experimentally synthesized and tested to possess superior optoelectronic properties compared to previous works, while simultaneously revealing design rules that govern the self-assembly of π-conjugated peptide systems.

625850 A Data-Driven Approach to Understanding and Predicting the Early Formation of the Solid-Liquid Electrolyte Interphase

Abstract:

A remaining challenge in the pursuit of rational design of novel and optimized electrolytes is understanding and ultimately predicting the reaction cascade responsible for the creation of a functional liquid-anode solid-electrolyte interfaces (SEI). We here present an overview of the challenge, in different energy storage systems and advocate that a data-driven approach coupled with a high-throughput quantum chemistry computational infrastructure[1] can address the complexity of such reaction cascade. To this end, we have constructed a chemical reaction network containing over 50,000 elementary reactions which allow for bond breaking/formation as well as oxidation/reduction reactions. The reaction energetics are obtained from our computational framework which automate geometry optimization and vibrational frequency calculations for molecules, including radical, charged, metal-coordinated, and solvated species. Our framework, using select levels of theory and a suite of on-the-fly error handlers, is able to successfully converge to a minimum energy structure and calculate the molecular thermodynamic properties in over 97% of cases. Individual calculations can then be combined to perform complex workflows in a fully automated fashion. To date, we have applied this framework to over 25,000 unique molecules relevant to the formation of solid-electrolyte interphases in Li-ion and Mg-ion batteries. To analyze possible reaction pathways, we apply a graph representation to our reaction network, leveraging an iterative reweighting strategy that solves for all prerequisite branching such that traditional pathfinding algorithms correctly capture coordinated pathways. Machine-learning algorithms operating on bond-formation and breaking energetics aid in rapid evaluation of highly reactive processes. We show that without any apriori chemical intuition, our automated framework recovers the most favorable reaction paths to form key SEI components, which were carefully identified over the past two decades. Thus, our data-driven approach and infrastructure show promise to accelerate the understanding of chemical reactivity in complex environments, in the aid of novel electrolyte design.

626136 Building a Universal Data Science Workflow for Monodentate Organophosphorus Ligands

Abstract:

The ability to forge difficult chemical bonds through catalysis has transformed society on all fronts, from feeding our ever-growing populations to increasing our life-expectancies by synthesizing new drugs. Not only has the rise in popularity of metal-catalyzed cross-coupling reactions enabled us to make existing processes more efficient, but it also has allowed us to synthesize novel and unexplored molecules and materials, unlocking the technologies of the future.

In metal catalysis, the choice of ligand often leads to the most significant impact on the reaction outcomes, such as yield or product selectivity. Identifying optimal metal-ligand combinations can be a laborious experimental process. This practice is often held back by the difficulty of meaningfully comparing results with different ligands. I will introduce our efforts to develop a platform for inverse design of catalysts utilizing high-throughput virtual screening (HTVS) and machine learning (ML), coupled with an extensive ligands database.

One of the most used strategies in ML for metal catalysis is the development of models based on reactant parameterization, pioneered by the Sigman group. Given that the space of potential ligands is immense, this tactic can be very challenging.

We have developed a platform aimed at data-driven ligand optimization and their inverse design. At the center of this strategy is Kraken, a database of descriptors for organophosphorus ligands that encompasses features that are most important for catalysis including conformational flexibility and ability for both coordinative and non-covalent bonding.

We characterize this chemical space systematically and demonstrate how these descriptors can be estimated for millions of ligands using ML, thus nearly completely mapping out an immensely relevant space of ligands.

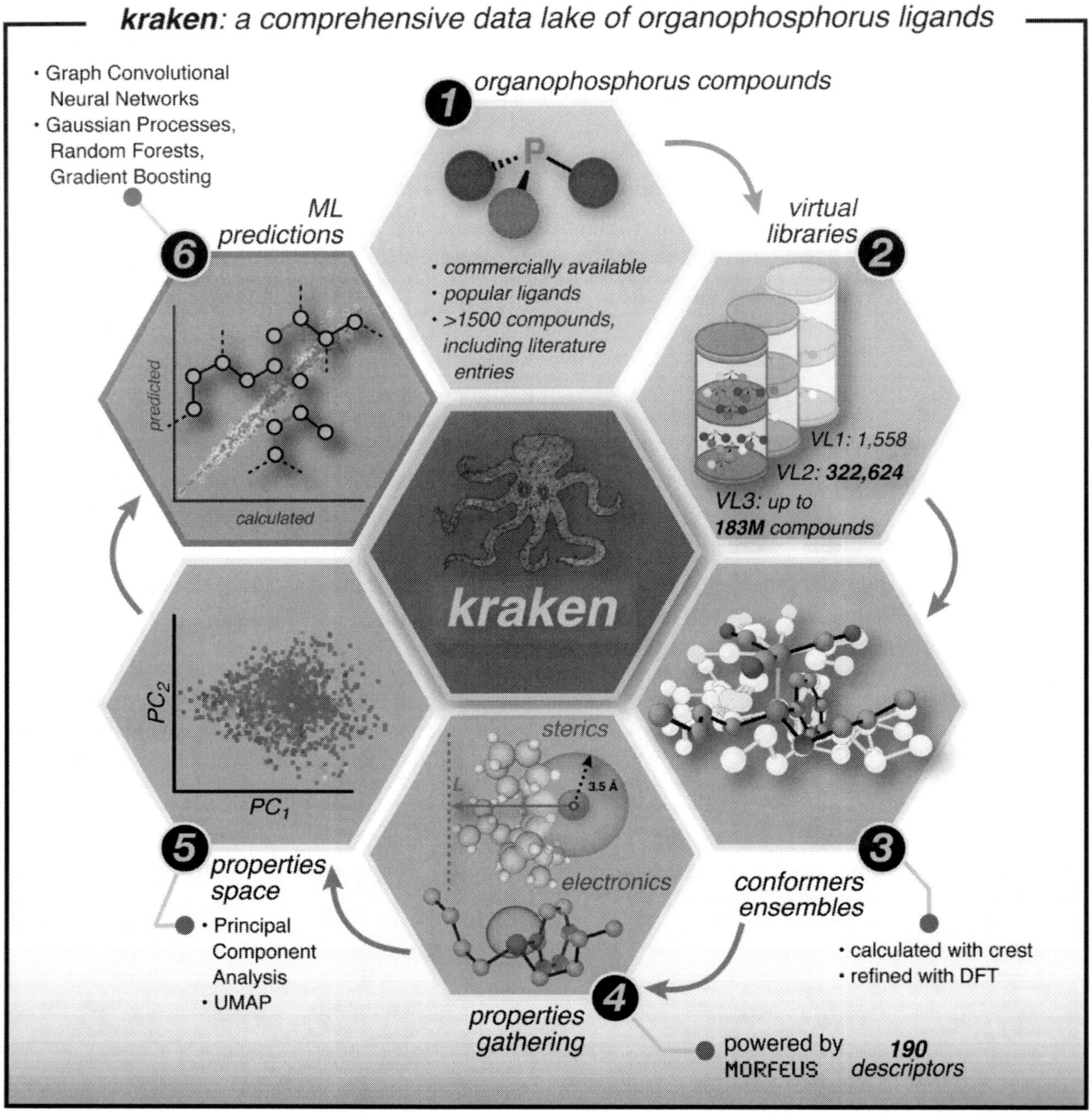
kraken: a comprehensive data lake of organophosphorus ligands
1 organophosphorus compounds
P
• commercially available
• popular ligands
• >1500 compounds, including literature entries
2 virtual libraries
VL1: 1,558
VL2: 322,624
VL3: up to 183M compounds
3 conformers ensembles
• calculated with crest
• refined with DFT
4 properties gathering
sterics
3.5 Å
L
electronics
powered by MORFEUS
190 descriptors
5 properties space
• Principal Component Analysis
• UMAP
PC_2
PC_1
6 ML predictions
predicted
calculated
• Graph Convolutional Neural Networks
• Gaussian Processes, Random Forests, Gradient Boosting
kraken

629873 From Brownian Dynamics Simulations and Experimental Observations of Colloidal Suspensions to Data-Driven Observables and Effective Sdes with Manifold Learning

Abstract:

We construct a reduced data-driven surrogate SDE model for electric-field mediated colloidal crystallization using data obtained from Brownian dynamics [1].
These simulations were previously matched to optical microscopy experiments on quasi-2D colloidal assembly in multipolar AC electric fields (that generate reconfigurable energy landscapes for charged colloidal particles[2-3])). Ongoing extensions of these experiments and simulations, also to be analyzed using the approaches reported in this work, include systematically varying particle shape and assembly on curved surfaces, which both introduce topological defects and new dynamics features.
To achieve this, we first simulate a system of 210 particles in the canonical ensemble under fixed voltage. The obtained configurations include fluid, polycrystalline and crystalline states [1]. The configurations are aligned to a reference configuration using the Kabsch algorithm [4-5] and then we use the manifold learning method Diffusion maps [6], with an appropriate metric, to learn the intrinsic geometry of the collected data set. Diffusion maps eigenvectors give a set of reduced/effective/latent variables in which we can learn the SDE. Those latent variables can also be tested for their physical explainability by utilization of the Inverse Function theorem in a data-assisted way with the use of the extension scheme Geometric Harmonics [7].
In order to construct an S.D.E. on our reduced/latent variables an effective gradient needs to be obtained. The identified drift and the diffusion coefficients provide a way to study the evolution on those latent coordinates through an effective Langevin equation [8]. We illustrate the estimation of the drift and diffusivity in the latent space in two ways; (a) a data-driven approach that allows us to approximate the coefficients using their statistical definition [8-9] (b) through a Neural Network architecture in which the state dependent drift and diffusivity are computed from paired snapshots (x_t, x_{t+h}), scattered over the domain of interest. This latter approach works pointwise and the loss function is formulated for Gaussian noise. Our reduced/effective trajectories are being "lifted" on the ambient space with the help of a Generative Adversarial Network (GANs). Lastly, we construct an effective potential based on the learned SDE.
[1] Yuguang Yang, Bevan, Michael A., Raghuram Thyagarajan, and David M.Ford. Dynamic colloidal assembly pathways via low dimensional models.Journal of Chemical Physics,144(20), 5 2016.
[2]. Zhang, Y. Zhang, M. A. Bevan, Spatially Varying Colloidal Phase Behavior on Multi-Dimensional Energy Landscapes. J. Chem. Phys. 152, 054905 (2020).
[3]. J. Zhang, J. Yang, Y. Zhang, M. A. Bevan, Controlling colloidal crystals via morphing energy landscapes and reinforcement learning. Science Advances 6, eabd6716 (2020).
[4] Kabsch W (1976) A solution for the best rotation to relate two sets of vectors. Acta Crystallogr A 32:922–933. 57.
[5] Kabsch W (1978) A discussion of the solution for the best rotation to relate two sets of vectors. Acta Crystallogr A 34:827–828.
[6] Ronald R Coifman and Stephane Lafon. Diffusion maps. Applied and Computational Harmonic Analysis, 21(1):5–30, 2006.
[7] Ronald R Coifman and Stephane Lafon. Geometric harmonics: A novel tool for multiscale out-of-sample extension of empirical functions. Applied and Computational Harmonic Analysis,21(1):31–52, 2006

[8] Ping Liu, CI Siettos, CW Gear, and IG Kevrekidis. Equation-free model reduction in agent-based computations: Coarse-grained bifurcation and variable-free rare event analysis. Mathematical Modelling of Natural Phenomena, 10(3):71–90, 2015
[9] H. Risken, T Frank ,The Fokker-Planck Equation: Methods of Solution and Applications Springer , 2012.

630026 A Data-Driven Approach to Disentangled Parametrization of Relations, with an Application to Multisite Phosphorylation

Abstract:

We developed and implemented a data-assisted procedure for parametrizing input-output relations [1]. Our suggested approach aims at parameter reduction in problems where the *effective inputs* are (possibly nonlinear) combinations of several original model inputs (parameters). Our data-driven approach is based on the manifold learning algorithm Diffusion Maps [2], the extension scheme Geometric Harmonics [3] and is assisted by deep learning (a conformal autoencoder neural network). The relations we considered in this work arise from coupled first order nonlinear differential equations describing the time-evolution of chemical/biochemical reaction networks. More precisely we considered two toy examples with three original inputs each and a model that describes the multisite phosphorylation with 6 original inputs total [4]. In our workflow, for each example, we collect a data set of output behaviors of the system and apply output informed Diffusion Maps. Diffusion Maps achieves parameter reduction and the new discovered machine learning effective parameters are given to us as eigenvectors of the Diffusion Maps algorithm. We establish in a data-driven way the physical interpretability of these "effective parameters" . These can be used to infer the parameters without optimization: for new, unobserved system behaviors we obtain the values of the data-driven and analytical effective parameters corresponding to those behaviors without running an optimizer. We also illustrate the ability of our scheme to map from new analytical effective parameters to systems' behaviors without the simulations. Utilizing our Y-shaped conformal autoencoder neural network allows us to go beyond discovering the effective paremeters, to parametrizing the level sets of the parameter combinations that do not affect the response. Lastly, we illustrate that Jointly Smooth Functions [5] provide another possibility to construct data-driven effective parameters.

[1] Alexander Holiday, Mahdi Kooshkbaghi, Juan M Bello-Rivas, C William Gear, Antonios Zagaris, and Ioannis G Kevrekidis. Manifold learning for parameter reduction. Journal of Computational Physics, 392:419–431, 2019

[2] Ronald R Coifman and Stephane Lafon. Diffusion maps. Applied and Computational Harmonic Analysis, 21(1):5–30, 2006.

[3] Ronald R Coifman and Stephane Lafon. Geometric harmonics: A novel tool for multiscale out-of-sample extension of empirical functions. Applied and Computational Harmonic Analysis,21(1):31–52, 2006

[4] Eyan Yeung, Sarah McFann, Lewis Marsh, Emilie Dufresne, Sarah Filippi, Heather A Harrington, Stanislav Y Shvartsman, and Martin Wuhr. Inference of multisite phosphorylationrate constants and their modulation by pathogenic mutations. Current Biology, 30(5):877–882,2020.

[5]Or Yair, Felix Dietrich, Rotem Mulayoff, Ronen Talmon, and Ioannis G. Kevrekidis. Spectraldiscovery of jointly smooth features for multimodal data.arXiv pre-print: 2004.04386, April2020

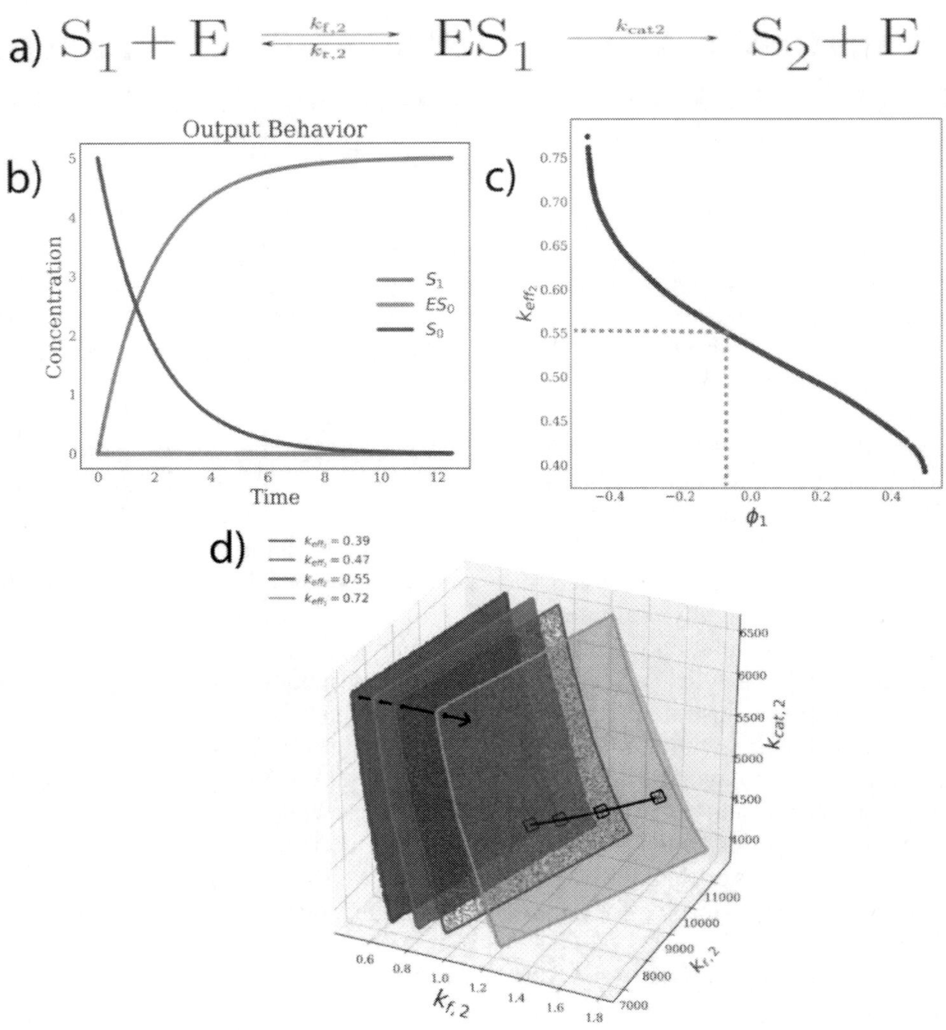

Figure: (a) The mechanism for which, if $k_{cat2}, k_{r,2} \gg k_{f,2}$, a single effective parameter $k_{eff_2} = E_0 \frac{k_{f,2}k_{cat,2}}{k_{r,2}+k_{cat,2}}$ describes fully the output behavior of the system. (b) The system's output behavior given a parameter combination $\boldsymbol{p} = (k_{f,2}, k_{r,2}, k_{cat,2})$. (c) Diffusion Maps coordinates ϕ_1 is one-to-one with the effective parameter of the model, (d) The level sets of the effective parameters, a single behavior of the system (red point in b) corresponds to a 3D level set of the original parameters (red curve in d).

631814 Predicting Chemical Adsorption Onto Montmorillonite Clays through a Combination of Experimental Isotherms, Minimalistic Simulations and Data-Driven Models

Abstract:

People and animals can be exposed to complex mixtures of hazardous environmental chemicals—such as polyaromatic hydrocarbons, pesticides, organic solvents, and toxic metals—through contaminated drinking water, food, and feed following natural disasters[1]. An attractive approach to minimize human and animal exposures to the toxic compounds is the use of edible, safe and effective sorbent materials to sequester them, thereby minimizing human and animal exposures. Previous studies and clinical trials have shown that montmorillonite clay significantly decreases biomarkers of aflatoxin exposure and could be safely consumed by humans and animals on a short-term basis[2]. In addition to aflatoxin, montmorillonite clay has been shown to tightly bind several other toxic compounds, including glyphosate and paraquat[3]. Due to its promise as a potential sorbent to mitigate chemical exposures, it is important to understand its function using a combination of computational and experimental methods[3,4,5], and develop new computational tools to screen and predict the efficacy of clay for different toxic chemicals.

In this study, we aimed to close the loop between experiments and computations, providing the potential for future experiments to be guided by computational predictions estimating the capacity of calcium montmorillonite clay to serve as sorbents for different toxic chemicals. We performed molecular dynamics (MD) simulations of montmorillonite clay in the presence of toxic compounds, and revealed key structural and physicochemical properties associated with chemicals that can be adsorbed to the clay. Using data derived from simulations and experiments, we developed a parametric, minimalistic model with the capacity to predict a compound's affinity to the clay with high accuracy. Thus, such a model can be used as a screening tool to predict and guide experiments on the efficacy of clay to bind other toxic compounds. Additionally, such predictive approaches, combining simulations and data-driven modeling, can be used to enable the design of clay-based sorbents to bind to toxic compounds of interest that are otherwise difficult to mitigate. This research is supported by NIEHS Superfund Research Program P42 ES027704.

1. Onel, M., Beykal, B., Wang, M., Grimm, F.A., Zhou, L., Wright, F.A., Phillips, T.D., Rusyn, I., Pistikopoulos, E.N., 2018. Optimal Chemical Grouping and Sorbent Material Design by Data Analysis, Modeling and Dimensionality Reduction Techniques. Aided Chem. Eng. 43, 421-426.

2. Phillips, T.D., Wang, M., Elmore, S.E., Hearon, S., Wang, J.S., 2019. NovaSil Clay for the Protection of Humans and Animals from Aflatoxins and Other Contaminants. Clays Clay Miner. 67, 99–110.

3. Wang, M. Orr, A.A., He, S., Dalaijamts, C., Chiu, W.A., Tamamis, P., Phillips, T.D., 2019. Montmorillonites Can Tightly Bind Glyphosate and Paraquat Reducing Toxin Exposures and Toxicity. ACS Omega. 4, 17702–17713.

4. Orr, A.A., He, S., Wang, M., Goodall, A., Hearon, S.E., Phillips, T.D., Tamamis, P., 2020. Insights into the Interactions of Bisphenol and Phthalate Compounds with Unamended and Carnitine-Amended Montmorillonite Clays. Comput. Chem. Eng. 143, 107063.

5. Wang, M., Orr, A.A., Jakubowski, J.M., Bird, K.E., Casey, C.M., Hearon, S.E., Tamamis, P., Phillips, T.D., 2021. Enhanced Adsorption of Per-and Polyfluoroalkyl Substances (PFAS) by Edible, Nutrient-Amended Montmorillonite Clays. Water Res. 188, 116534.

621660 Machine Learning Electronic Structure for Heterogeneous Catalysis

Abstract:

The electronic structure of a material plays a key role in its surface chemistry and catalytic performance, but the process of utilizing this information for quantitative predictions is often challenging. One such approach is to develop descriptors, such as the d-band center, which condenses this information into individual features which are more amenable for property prediction. However, identifying valid descriptors via human intuition can be time consuming and quickly becomes intractable for chemically diverse systems, thus necessitating a more flexible and automated approach. Here we describe how machine learning can be used to obtain the electronic structure features via a trainable process, which can furthermore efficiently leverage existing high-throughput datasets for catalysis. We find this approach can easily outperform conventional descriptors in prediction accuracy while maintaining a high degree of physical meaning and interpretability in the models. The electronic structure space can consequently be effectively accessed and explored with significant implications for catalyst discovery and design.

622602 Machine Learning the Quantum-Chemical Properties of Metal-Organic Frameworks for Accelerated Materials Discovery with a New Electronic Structure Database

Abstract:

Metal–organic frameworks (MOFs) are a widely investigated class of extended solids with tunable structures that make it possible to impart specific chemical functionality tailored for a given application. However, the enormous number of possible MOFs that can be synthesized makes it difficult to determine which materials would be most promising, especially for applications governed by electronic structure properties that are often computationally demanding to simulate and time-consuming to probe experimentally. Here, we present the first publicly available quantum-chemical database for MOFs (the "QMOF database"), which consists of properties derived from density functional theory (DFT) calculations for over 14,000 experimentally synthesized MOFs. Throughout this talk, we demonstrate how the QMOF database can be used to identify MOFs with targeted electronic structure properties. As a proof-of-concept, we use this database to evaluate the performance of several machine learning models for the prediction of DFT-computed band gaps and find that crystal graph convolutional neural networks are capable of achieving high predictive performance, making it possible to circumvent otherwise computationally expensive calculations. By using the QMOF database and resulting machine learning models, we were able to predictively identify several MOFs with low band gaps, a challenging task given the electronically insulating nature of most MOF structures. We conclude by highlighting recent and forthcoming updates to the QMOF database that will further expand its ability to accelerate the discovery of promising MOFs with targeted electronic structure properties.

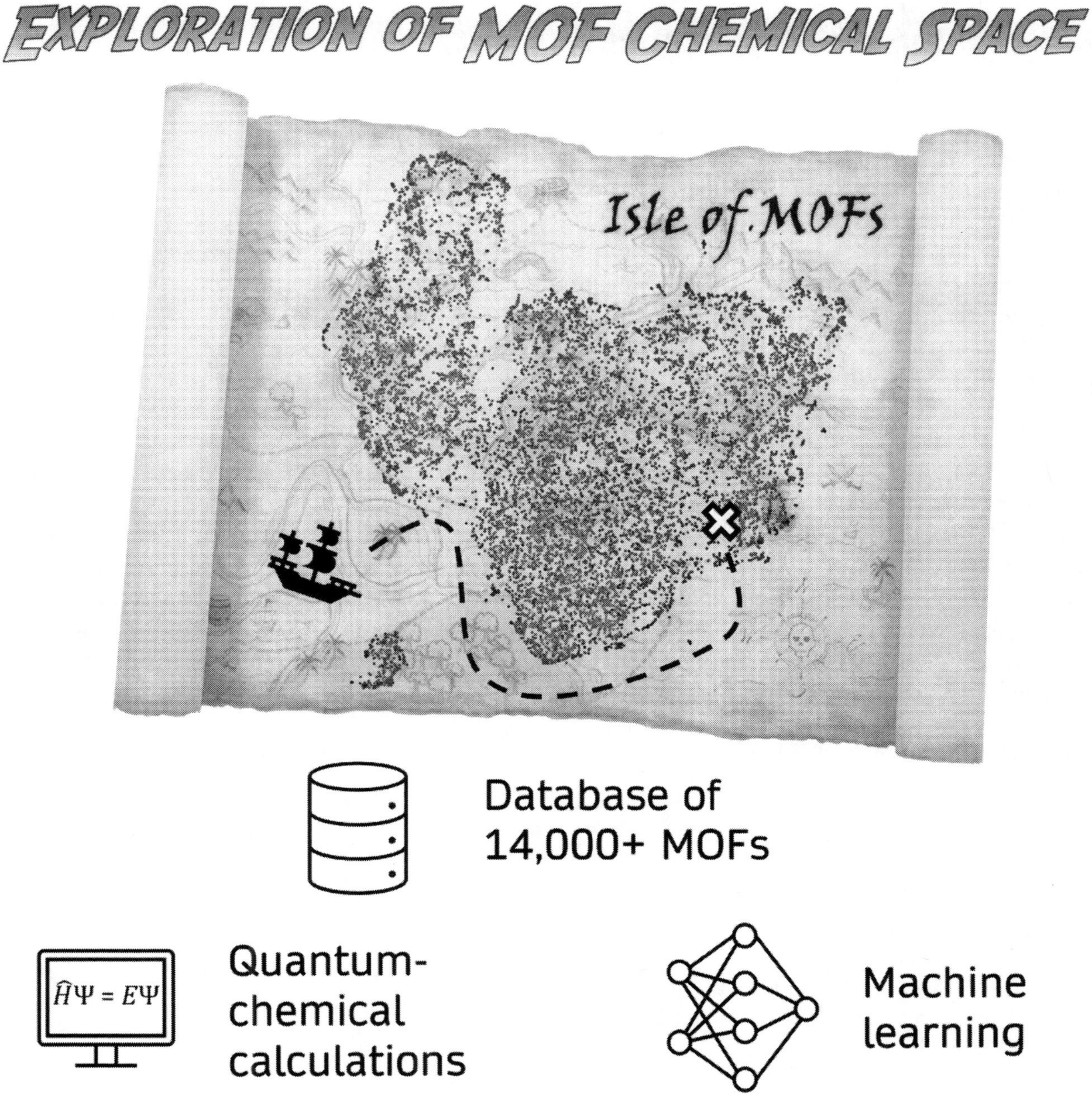
EXPLORATION OF MOF CHEMICAL SPACE
Isle of MOFs
Database of
14,000+ MOFs
$\hat{H}\Psi = E\Psi$
Quantum-
chemical
calculations
Machine
learning

624176 Bayesian Parameter Estimation for Connecting Theory & Predictions to Experiments

Abstract:

A frequent goal of research is to find the correct physical parameters for chemicals and materials (energetics, lattice constant, etc.). Bayesian parameter estimation (BPE) is intended to give the most probable set of parameters for a given model and a given data set. However, this same methodology can be applied to multiple types of problems in different ways. Broadly speaking, application of BPE can be divided into two categories. One way, the more 'modern' usage of BPE is obtaining parameters which are adequate for creating predictive surrogates (this is often what BPE is used for in machine learning, and is the philosophy behind much of the application of Gaussian Processes to sparse data sets). This first category can be considered analogous to making a calibration curve with fitting. A second way, and more 'traditional' application of BPE is for finding the parameters which have the highest probability of being correct, for a given model. Additionally, this second category of BPE can also be used for model selection (which model is more probable given the data). This second category of BPE is particularly useful for comparing theoretical predictions to experiment. Importantly, practitioners using the first category of BPE to make theoretical predictions can then use the second category of BPE to compare with experiment. In this talk, the usefulness of this concept will be presented, as well as how it is implemented in the free, open source, and user friendly CheKiPEUQ software. This software was developed with U.S. Department of Energy funding, and has been designed for science and engineering applications.

625189 Towards Development of Novel Heterogeneous Catalysts Using Extrapolative Machine Learning Methods

Abstract:

Molecular/materials informatics has become a central paradigm in molecular and materials science thanks to the enormous potential it holds to revolutionize the design of functional molecules/materials. However, although we have already seen proof-of-concept examples that artificial intelligence (AI) can reduce the time and costs involved and also can find new compounds, most of them have only been tested on benchmark problems and no fundamentally-new molecules/materials or synthetic-transformations have been found. This is primarily due to lack of data and that machine learning (ML), the main player in this campaign, is highly focused on optimization rather than finding novel compounds and phenomena (extrapolation).

It should also be mentioned that establishing "Catalysis Informatics" is even more challenging. Although it is highly related to materials informatics and chemoinformatics, it is distinguished by the fact that catalysis is a time-dependent dynamic event controlled by the structures and chemical nature of catalytically active sites. In particular, heterogeneous catalysis is still a largely empirical science due to the complexity of the surface chemistry involved. This situation causes lack of data as the computational costs to obtain accurate theoretical models for such complex heterogeneous catalysis are currently prohibitively high and high-throughput experimental methods, which have been applied successfully to relevant fields, have not been explored fully at the current time. In this regard, building ML models that effectively find novel catalysts within diverse chemical space from "real world" experimental catalysis data (not from well-behaved computational data) is highly desirable.

In this context, we have proposed a ML approach which uses elemental features as the input representations rather than inputting the catalyst compositions directly. Namely, in our proposed method, the elemental composition ratios are multiplied by elemental descriptors such as electronegativities, melting points, atomic radii, etc. which are unique for each element. We call this approach the Sorted Weighted Elemental Descriptor (SWED) representation. Importantly, this new ML method has the potential to guide catalyst design and discovery in areas where limited catalyst composition overlap exists and even for elements previously unseen in the given data, enabling us extrapolative and ambitious exploration beyond the training data. We have used the developed ML approach to analyze literature data on oxidative coupling of methane (OCM) and water gas shift (WGS) reactions. The ML method was found to be effective for predicting novel promising catalyst candidates that include elements unseen in the original dataset for future studies (Figure 1).

It should also be noted that analysis using the extrapolative ML can reveal not only effective catalyst compositions but also the required elemental features and electronic properties so that ideal catalysts can be designed in a highly precise manner. Because catalytic properties of materials in principle should be determined by their electronic structures, the strategy is to design target electronic structures by changing the composition and physical nature of selected materials. The concept of controlling the properties of matter at the molecular scale by engineering electronic structure should not only be relevant to catalytic materials but also more generally applicable to other challenges in materials science.

(a) the number of each elements appeared in the original WGS dataset

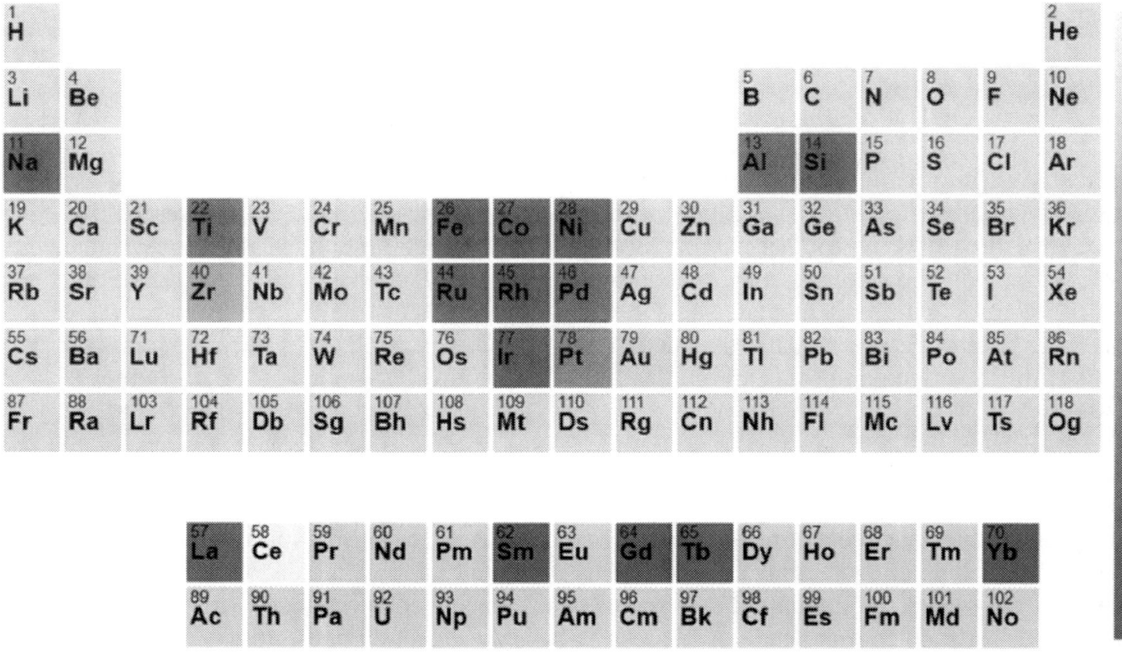

(b) the number of each elements appeared in the ML-predicted catalysts

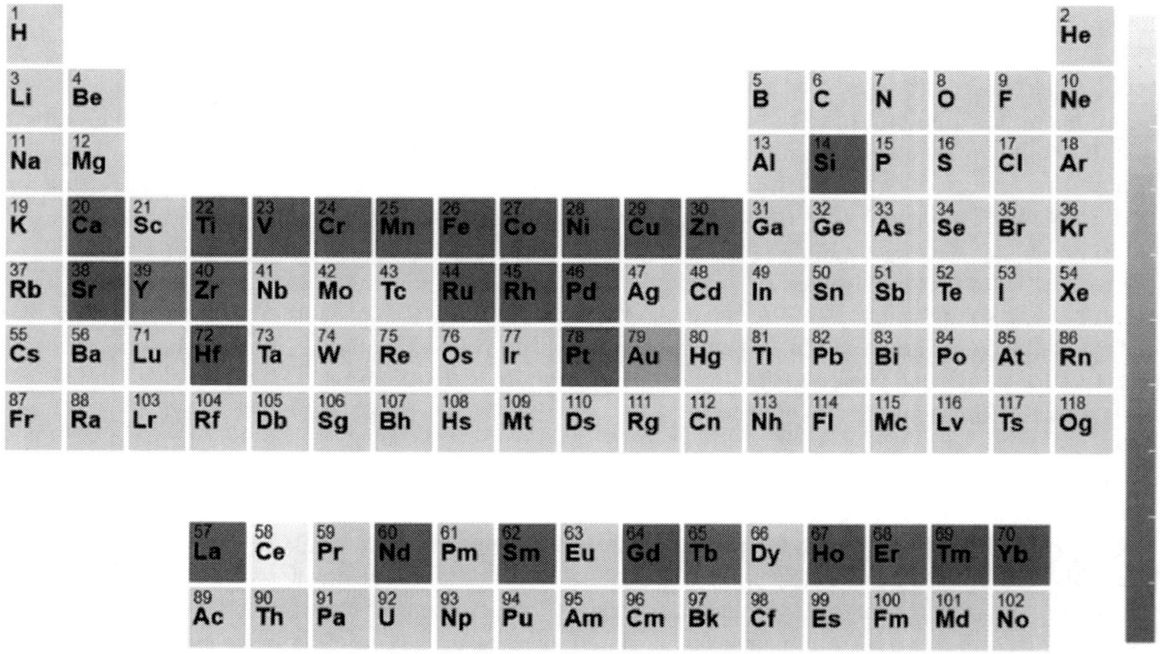

AUTHOR INDEX

AIChE
120 Wall Street, FL 23
New York, NY 10005-4020

ISBN 978-1-7138-5689-4

CURRAN ASSOCIATES INC.
proceedings.com